The American Street Gang

STUDIES IN CRIME AND PUBLIC POLICY

Michael Tonry and Norval Morris, *General Editors*

Police for the Future
David H. Bayley

Incapacitation: Penal Confinement and the Restraint of Crime
Franklin E. Zimring and Gordon Hawkins

The American Street Gang:
Its Nature, Prevalence, and Control
Malcolm W. Klein

The American Street Gang

Its Nature, Prevalence, and Control

Malcolm W. Klein

New York Oxford
OXFORD UNIVERSITY PRESS
1995

Oxford University Press

Oxford   New York
Athens   Auckland   Bangkok   Bombay
Calcutta   Cape Town   Dar es Salaam   Delhi
Florence   Hong Kong   Istanbul   Karachi
Kuala Lumpur   Madras   Madrid   Melbourne
Mexico City   Nairobi   Paris   Singapore
Taipei   Tokyo   Toronto

and associated companies in
Berlin   Ibadan

Published by Oxford University Press, Inc.,
198 Madison Avenue, New York, New York 10016

Oxford is a registered trademark of Oxford University Press

Library of Congress Cataloging-in-Publication Data
Klein, Malcolm W.
The American street gang : its nature, prevalence, and
control / Malcolm Klein.
p.   cm.—(Studies in crime and public policy)
Includes bibliographical references and index.
ISBN 0-19-509534-0
1. Gangs—United States.
2. Juvenile delinquency—United States—Prevention.
I. Title. II. Series.
HV6439.U5K58 1995
364.1'06'0973–dc20   94-31402

1 3 5 7 9 8 6 4 2
Printed in the United States of America
on acid-free paper

For the sustaining forces in my personal life:
In the early years: Terry, Laurie, and Leigh
In the later years: Margy

Abstract

The American Street Gang reviews what has been known about gangs and updates that information into the 1990s. It covers reported changes in the structure and crime patterns of gangs, their age, ethnic and gender characteristics, and their spread into almost all corners of the nation. It also reviews and updates the situation in other countries to determine whether the American gang is unique.

This is not a textbook. The author has a number of things to say to the reader that are based on decades of research and personal experience and that contradict many popular and professional beliefs about street gangs. The early chapters spell out four issues that are major concerns in understanding the gang problem and then weave these four issues throughout the book.

The first is the issue of defining the gang, accomplished here by describing street gangs and separating them from other forms of gangs, such as skinheads, peer groups that occasionally engage in delinquency, drug distribution gangs, and motorcycle gangs. The author sees the street gangs as qualitatively different from these other groups, though still having many structural variations.

The second issue involves recent changes in gangs, including their violence and their spread to eight hundred towns and cities in the United States. Brand-new data are presented here for the first time.

The third issue is street gangs' involvement in drug distribution. The author comes down very hard on those who have equated gangs and drug sales. He presents his own data, along with the findings of other researchers, to provide a more balanced picture than is often portrayed in the media.

The fourth issue, ways of handling the gang problem, is equally controversial. The author reviews gang prevention and suppression programs and suggests that most have been either ineffectual or actually have made matters worse. Practitioners will not like the message here but would do well to consider the author's conclusions.

The American gang situation is worsening, and our response is not improving. The author's thoughts on this situation will stir some strong reactions.

Preface

I started doing gang research in the autumn of 1962 and have now been at it, off and on, for some thirty years. Starting in total ignorance, I have accumulated enough exposure, knowledge, frustration, anger, and recovery to attempt to write a book that might help move the field a bit. The journey has been guided at various points by friends and colleagues. All of the former have been sources of knowledge and support; of the latter, some have been supportive, and some have provided the opposition needed to focus and sharpen my views. I need to acknowledge the good guys and the others, explicitly.

LaMar Empey, eminent criminologist, guiltless big game hunter, and occasional role model has probably done more to mold my career than any other professional. I like the mold, and readers of this volume will unknowingly be affected by it.

Cheryl Maxson, my gang research colleague over the past dozen years, has become my methodological superego and conceptual partner. Little that I say in this book is mine alone, but she would find nicer ways to say it. So would Lea Cunningham, our long-suffering and absolutely invaluable field supervisor.

Elaine Corry is one of the most amazing people I've ever known. Friend, colleague, administrator, tolerator of all sorts of human frailty in others, she has been for me and all the others in our research institute our most indispensable resource. She talks now about retirement; if she does it, we may all fall apart like the one-hoss shay.

In the world of gangs and cops, one man has done more, and more willingly, than any other to provide access to the world of the gang cop. Sergeant Wesley McBride of the Los Angeles Sheriff's Department (LASD) started his gang intelligence work back in the early 1970s. He's still at it, having become the most reasoned and knowledgeable gang cop I know. My acknowledgment of his help should be taken, as well, as recognition of the nondefensive, administratively open stance taken by the Sheriff's Department toward my many intrusions into their world.

I have benefited from competition and opposition as well. I wouldn't have carried out street gang research in Los Angeles over these many years without the approval of former chiefs Parker, Davis, and Gates of the Los Angeles Police Department (LAPD). I've been the LAPD's buddy—cited, praised, consulted on those occasions when my data and views supported the department's position on gangs. I've been the LAPD's punching bag—belittled, attacked, and ignored as the occasion dictated when my data and views were not ideologically acceptable. The LAPD's approach to gang intervention over the years will receive attention in this volume.

I should add, briefly, that the media's portrayals of gang matters, often taken from police with their own specialized vision, have also served to sharpen my views. I spend many hours each month responding to TV, radio, and print reporters, trying to divest them of their assumptions and alerting them to the accumulated social science literature on gangs. Sometimes I succeed; more often, I do not.

Let me add that some of my social science colleagues are on occasion the purveyors of false gang images. Sometimes they manage this out of ignorance, repeating the media's images of gangs rather than looking at their colleagues' research. Sometimes they go off base by leaning uncritically on idiosyncratic studies.

On the positive side, I need to acknowledge that most academic colleagues have listened carefully to my counterintuitive views on gangs; that a number of working staffers in local and federal agencies have successfully divorced their agency's ideology from their appreciation of empirical research; that some media representatives have developed an expertise on gang matters that is reflected in their coverage of complex issues; and that many enforcement officials across the nation have broken from the narrow deterrent mold and kept an open mind on gang issues.

Other names need to be mentioned. Over a period of many years I have listened most closely to three gang scholars: James F. Short, the dean of gang researchers and one of the few to appreciate the importance of group process in gang behavior; Irving Spergel, whose decades of gang research have retained the focus of community structures in understanding gang existence; and Solomon Kobrin, my informant on the weird world of Chicago, as well as working colleague, decent human being, and true scholar in the best sense of that word.

Others in criminology who may be surprised to learn that they have significantly affected my thinking are Delbert Elliott, Jeffrey Fagan, Marcus Felson, Daniel Glaser, and John Hagedorn. Each contributes a viewpoint, a methodological flavor, and a respect for data that I carry with me and that flows through my pen to the yellow legal pad on which I try my thoughts.

Also, I want to express my thanks to a number of people who have helped me connect with the gang situation elsewhere in the world. Their contributions will be in evidence when I discuss gang issues in Spain (Rosemary Barberet and Cristina Rechea Alberola), England (David Farrington, Betsy Stanko, Dave Robins, Mike Chatterton, Ken Pease, Divisional Narcotics Inspector Charles Coxon and the command staff at the Moss Side Station in Manchester, and Detective Inspector Martin Challis of the NCIS in London), Papua New Guinea (Richard Sundeen), Australia (Kenneth Polk), Germany (Walter Specht, Hans-Jurgen Kerner, K. O. K. Manfred Bauer of the Frankfurt police, Claudius Ohder and Alexis Aronowitz), medieval Japan (Peter Nosco), Russia (Elena Savinkova), Switzerland (Martin Killias), and Sweden (Jerzy Sarnecki, Per-Olof Wikstrom, Anne-Marie Janson, journalist Jonas Hallen and photographer Alexander Ruas, and a cop who prefers anonymity). I am limited in my language skills and therefore gratefully acknowledge a good deal of translating done for me by Victor Agadjanian (Russian), Jurgen Wehner and Bob Roberts (German), Eli Heitz (French), and Letty Baz (Spanish). The English, I fear, can be blamed only on my own teachers.

This manuscript has been written at home, in my office in Los Angeles, and in an office generously supplied by the Swedish Council on Crime Prevention during my sabbatical leave in Stockholm. I visited other European cities with support from a University of Southern California Zumberge Research and Innovation Fellowship. Judith Webb somehow managed to coordinate all the mailings and faxes and shipments to produce a single document. She has also dotted my i's, crossed my t's, and gently reminded me of what her teachers said, as well as mine. Our combined efforts have been reviewed in draft form by some long-suffering friends and colleagues—Margy Gatz, Cheryl Maxson, James F. Short, Irving Spergel, Sergeant Wes McBride (Los Angeles Sheriff's Department), Chief Lorne Kramer (Colorado Springs Police Department), and Michael Genelin (Los Angeles District Attorney's Office)—whose advice I have alternately accepted, rejected, modified, and in others ways incorporated so that I alone can take blame for the results. My thanks to them, to Judith, and to the Swedes for facilitating the enterprise.

My final thanks go to George Ritzer whose idea this was to start with, allowing me the opportunity to write an uncharacteristically personalized account of some thirty years in street gang research.

Los Angeles
September 1994   M. W. K.

Contents

The American Street Gang

1

Introduction

After six months, you'll know all there is to know.

Hans Mattick, winter 1962

Hans Mattick was wrong. Not in six months, or in thirty years, have I learned all there is to know about gangs. But what I have learned, I now want to share.

This book is about the American street gang. Despite its very considerable variety—over time, across locations, and even in its internal substructures—I see the American street gang as rather different. That it is not totally different is also clear: Although I'll describe counterparts in such contrasting places as Russia, Papua New Guinea, and Berlin, the common varieties of street gang still are essentially an American product.

We need to understand this gang for intellectual purposes, to seek to control its excesses and prevent its continuing regeneration, and also to understand our society better. Gangs are no accident; our society inadvertently produces them, and they will not decline as a social problem until we confront our relationship to them. And to confront our relationship to street gangs is to come face-to-face with some well-entrenched self-interests that also are important to understanding ourselves. Gangs have a societal context, and to paraphrase Pogo, the context is us.

How can we know a gang when we see it? Does a gang member stand out in some way? How do we distinguish a gang-related criminal act from any other? As a case in point, consider the most infamous incident in the Los Angeles riots in 1992, occasioned by the acquittal of the police officers who beat black motorist Rodney King to a pulp. It was the "payback" beating of a white trucker, Reginald

3

Denny, seen on news broadcasts as it happened, that came to epito-
mize the lawlessness of the riot.

Four men were arrested for the attack on Denny, who barely sur-
vived the attack. Later, additional arrests were announced in connection with other beatings close in time and place to the Denny assault.
From the moment of the first broadcast of these events—most involving black assailants on white, Asian, and Hispanic victims—gangs
were blamed. Gangs were beating up innocent bystanders; gang
members were leading and directing the assaults, even selecting the
victims; the looting and arson had become a gang event. So said the
broadcasters, so parroted the newspapers, and so assumed much of
the shocked and frightened population of Los Angeles. The United
States was exposed to its first gang rebellion.

A gang name was offered—the Eight-Trey Gangster Crips—and
the phone calls started coming in to me:

"Is there such a gang?" Yes.

"Are these suspects really gangbangers?" I don't have any way of
knowing.

"Are gangs behind the rioting?" I don't have any way of knowing.

"Doesn't the viciousness of the beatings prove this is a gang
event?" No; gangs have no monopoly on violence.

As this is written, the court process has just been completed, with
several convictions for relatively minor offenses. The prosecution was
pushing the gang connection because it would lead to sentence
enhancements under new California law—up to life in prison in some
cases. Two of the four defendants had been identified as gang members by a Los Angeles Police Department (LAPD) gang unit officer.[1]
The defense was crying "foul," a media event preventing a fair and
just trial. Some local residents, ignoring their usual status as potential
gang victims, were demonstrating for release of the "Los Angeles
Four," and the prosecution was pushing the drug and hoodlum
theme. And no one, as far as I can determine, was attempting to dis-
entangle the definitional issues—what a gang event is. No one openly
attempted to understand the meaning of the event in the context of
what we know about gangs.

Major Themes

The chapter headings of this book provide a snapshot of some of this
context: persistent gang issues, historical background, the contemporary picture, approaches to gang intervention and control, and
prospects for the future. Woven through all of these materials are four
central issues that are interdependent but require separate treatment
in Chapter 2.

I need to discuss the definitional issues—the first of the four—
with enough care that the reader can share with me a reasonably

consistent image of the American street gang, understanding what is and is not included in that image. Just how should "street gang" be defined? We're talking about some terminology arbitrarily, ambiguously, and variably applied to a shadowy, shifting, and diverse phenomenon. Media images, which form so many of our views, are usually inconsistent distortions of street gangs. We have to get beyond them.

We also have to get beyond simplistic notions of the street gang. As an example, there's an audiotape that I saw advertised in a law enforcement newsletter. It is probably one of many audio- and video-tapes now available, because the country is being criss-crossed by traveling gang road shows. Many groups, but most notably the Los Angeles Police and Sheriff's Departments (LAPD and LASD), the Chicago Police Department, and the National Center for State and Local Law Enforcement Training, funded by the U.S. Department of Justice, offer seminars and workshops on gang matters.[2] You can attend these workshops, usually aimed at police, parole, and other criminal justice personnel, and learn to recognize gang colors, hand signals, graffiti, and verbal argot; get exposure to gang values and behavior patterns, including ethnic variations; receive training in investigative techniques special to gang cases; and often come away with an official, signed certificate verifying your workshop attendance.

The audiotape, which I purchased, features the LASD's gang experts passing on their lore to a very appreciative police audience in the east. At one point, an LASD captain, long known in California for his understanding of street gangs, is recorded on the tape discussing the proper definition of gangs. He quotes a slightly long definitional statement written some years ago by a professor of sociology and deliberately trips over the long words—*denotable, aggregation, delinquent incidents*—to the audible glee of his audience. Then the captain offers his own definition: "I'll tell you what a gang is; it's a group of thugs. They're hoodlums, they're crooks and criminals."[3]

The quoted professor, you may have guessed, was me, so you can imagine my amusement at first hearing my own work as the butt of the good captain's humor. But you can also appreciate that the captain's engaging definition is no more enlightening than the gang member's response the first time I asked one of them what a gang was: "We a defensive club, man."

To move the dialogue on to a more useful level, I'll provide a depiction—not a definition—of street gangs in the next chapter sufficient to set some parameters for the three other central issues: gang proliferation, the gang–drug connection, and the place of group process in gang control.

The second central issue is gang proliferation, the appearance (but not necessarily the spread) of gangs in more than eight hundred

American towns and cities. Of all the changes that have taken place in the gang world over these past thirty years—the easy availability of firearms, ethnic changes, age changes, violence, drug involvement—none stands out more than the simple fact of the presence of street gangs in so many communities. A stable definitional approach allows us to compare what was with what is. This comparison in turn raises the etiology question—"How come?" Coverage of the definitional, proliferation, and "how come?" issues bring us to the discussion of my favorite gang variable, cohesiveness, and to issues of gang control.

The third issue I've chosen to stress is the relationship between gangs and drugs. This issue was not chosen because it is as important as definitions, proliferation, or control—it is not—but because it is the hottest gang topic in the press, among many police officials, and especially in the federal agencies concerned with gang issues. As a foretaste of what's to come, I'll cite just one illustration, an unfortunate anecdote that points up some of the problems I'll be alluding to.

After a lengthy investigation, the Federal Bureau of Investigation (FBI) announced in early 1992 the arrest of a half-dozen persons and the seizure of a very large amount of cocaine. This happened in Los Angeles. The announcement, dutifully reported by the press, stated that this major event proved the direct connection between the Los Angeles street gangs and the Medellín cartel. The gang involved was stated to be the "Muslim Crips," and six arrestees were named.

As it happens, the Los Angeles Police and Sheriff's Departments had been gathering intelligence on gangs for the past several decades and so could produce at this point an extensive list of names and identification data. As soon as the FBI bust was announced, both local intelligence teams searched their files and computerized rosters of a thousand gangs and more than 100,000 gang members to locate this cartel-connected street gang. But they found no indication of a "Muslim Crips" gang and no mention in the computer system of any of the six gang arrestees. The FBI had indeed broken up a drug sales ring, but not a street gang connected to the Colombian cartel.

Local law enforcement officials often question the expertise of FBI agents when they are working the same territory. In this instance, the local officials informed me of this error in the context of the recent pledge to release three hundred FBI agents from their cold war assignments to assist in the war on gangs in major U.S. cities.

Mind you, I do not cite this minor incident to poke fun at the FBI—it does not have trained gang officers—but, rather, to point up the gap between street gang realities and the image of them that has taken hold of many officials in the federal and state enforcement world. I could as easily have mentioned Operation Streetsweep, in which agents of the Bureau of Alcohol, Tobacco, and Firearms (ATF) began a secret mission in Los Angeles by flying surreptitiously in an Air Force C-5A jet transport into El Toro Marine Base in Orange

County, lest they be detected, I suppose, by the street gangs who have secretly infiltrated air traffic control at Los Angeles International Airport!

Or I could have cited the Drug Enforcement Agency's report that one can easily demonstrate that LA's Crips and Bloods have infiltrated and franchised the drug markets in many major cities across the nation, simply by looking at a map showing that all of these cities are connected to Los Angeles by major interstate highways. Can someone name a major U.S. city that is not connected to our interstate highway system?

Nor do state officials escape. I could have turned to the California State Task Force on Gangs and Drugs that so creatively, and ineptly, overstated the situation as follows:

> Today's gangs are urban terrorists. Heavily armed and more violent than ever before, they are quick to use terror and intimidation to seize and protect their share of the lucrative drug market. Gang members are turning our streets and neighborhoods into war zones, where it takes an act of courage simply to walk to the corner store.[4]

Obviously, my four issues overlap. These examples point up the definitional issue, as well as the implications of the proliferation of gangs into a national problem calling forth a response from a national (federal) agency. Clearly the gang connection to drugs is of concern here. In some ways, my colleagues at the University of Southern California and I have become the whipping boys (and girls) for the enforcement world because we were the first to question that gangs were heavily involved in the distribution of crack. "Rock" cocaine or "crack" was first seen in Los Angeles and first seen by the LAPD. The year was 1982, and for a while the department was puzzled about crack and its implications. We soon became involved in a way that has shaped our relationship with the police ever since. I'll describe this saga in later pages.

The fourth and last central issue threading its way through this book is that of gang cohesiveness, a concept that encompasses many variables describing group process and structure. Gangs are groups, after all, no matter how differently people may define them. Gang scholars argue about whether gangs are or are not cohesive groups. They do so, mind you, without bothering to measure gang cohesiveness. Consider the following:

Most social science literature indicates that increasing group cohesiveness also increases group morale and productivity. One of the products of gangs is crime.

Most gang intervention programs, upon analysis, can be shown (have been shown, can be inferred to show) an increase in gang cohesiveness.

The obvious implication is that most gang intervention programs, without meaning to, have the net effect of increasing gang crime.

If this little syllogism has any validity—and I'll attempt to demonstrate just that—then we had better be very careful about our gang interventions. This is a warning to social workers and cops alike and to those public officials who allow, support, and encourage their anti-gang policies and programs. Antigang activities may inadvertently promote gang crime and violence. Caveat emptor.

Levels of Understanding

In order to weave these themes through the chapters to follow, I need to explain my procedure, as it departs somewhat from the usual style expected by my academic colleagues. It will seem less cautious, less copiously footnoted, and less nonprescriptive than has been my style in the past. The approach is much like that described by Jim Short in his 1990 survey of the delinquency field: "In doing this, the book is selective, rather than exhaustive; focused, rather than encyclopedic."[5]

Just as gangs have proliferated across the nation, so has gang literature. We've come to the point that extensive reviews of the gang literature are at hand, freeing me from the need to undertake such a review in this volume. There are even textbooks on gangs now, suitable for a whole semester's class on the topic.[6] Indeed, I know of several colleagues who now offer a course on gangs at their universities.

There also are a number of interesting research monographs and books concentrating on particular gangs or particular regions that comprise a new generation of gang information. Readers may wish to look at John Hagedorn's *People and Folks*, Joan Moore's *Going Down to the Barrio*, Diego Vigil's *Barrio Gangs*, Luis Rodriguez's *Always Running*, and Mercer Sullivan's *Getting Paid*. I have some reservations about Felix Padilla's *The Gang as an American Enterprise* and Leon Bing's journalistic essay *Do or Die*, but both are good reads, as is Bob Sipchen's documentary novel *Baby Insane and the Buddha*. For readers seeking more of a shotgun approach, C. Ronald Huff's edited volume, *Gangs in America*, contains a number of papers that exemplify the directions that gang theory and research are now taking.

With all these works, not to mention the many articles on gangs now appearing in the professional journals, I really have few qualms about abandoning the usual academic monograph style. The situation allows me to break loose and be a bit more informal than has been my wont. I don't mean to overdramatize my newfound freedom or to set up the reader for more than I can produce. The point is that I will summarize the literature on various points without meticulously referring to every source, and I will draw conclusions and make suggestions that in many instances come from my own experience.

In order to do this, I need to relate to the reader my own involvements over the past thirty years. My views are based on personal observations, carefully collected data, long-term study of other people's research, hundreds of conversations with gang members and cops and social workers and reporters and politicians and bureaucrats and other researchers, and so on. Also, like any good academic introvert, I've discussed my views over and over with myself, an odd process that, like teaching, serves to reveal one's own, less well articulated thoughts. I know what I think by listening to myself.

I started as a neophyte, thrown into the pool to see whether I could swim. A four-year project to evaluate a gang intervention program floundered in its first year. The research team, headed by a half-time director, was not relating well to a resistant and suspicious team of gang workers and their supervisors. I was asked to come on as a full-time research director and "make things happen" (sink or swim, in fact). I was hesitant until they took me to my first gang meeting. I need to describe the scene—not dramatic but very revealing.

The place was the recreation hall of a large local park known as a hangout for the Operators.[7] In a crowded, hot meeting room, about a dozen youths were seated formally around the table. I judged their ages to range from perhaps 14 to 20 or so—at that point, I'd had no need to learn how to judge age. They were dressed in slacks or levis, white or plaid shirts, and sweaters despite the heat. Many wore "stingy-brim" hats. They weren't what I was used to—among other differences, the kids were black—but they were calm and only a bit fidgety. That would change.

At the head of the table sat an older youth in a suit: "Four-eyes" was the president of the "club" and the only member who wore glasses. Monikers, I learned later, were often based on an individual weakness or idiosyncracy, not a strength. "Crazy Benito" and "Potato Face" and others like them were often statements of fact. "Four-eyes" wore glasses because he was attending classes at a community college; he also came equipped with a briefcase.

Slightly behind and to the left of Four-eyes was Jacob, the worker just recently assigned to the Operators. It was his first formal meeting with them, and I was his guest. I sat at the foot of the table, but two yards back, at the proper "observer distance." I was also closest to the one door, by my own design (how did I know what might happen at a gang meeting?). And around the outside of the table sitters sat another half-dozen youths. I presumed they wanted to be close to the breezes occasionally wafting in through the open western windows.

The conversation was a bit hard to follow. Orderly presentation was clearly not the norm; there were references to people and events unknown to me; and also it was my first introduction to the dialect of the black gang member. Jacob, it seems, was discussing his role and using Four-eyes as his source of legitimacy. It seemed clear that they had conferred at some earlier point. Jacob had been assigned to the Operators, whereas his predecessor had been a volunteer, a local adult who hung around the park and connected with the Operators as a freelance "sponsor" (Oh, how the police hated the gangs' use of the word *sponsor* to describe their gang workers; "My god, Doc, how can you go out and deliberately sponsor these gangs?").

Slowly but surely—and quite deftly, I thought—Jacob was able to elicit the group's misgivings about the previous worker and to provide his own qualifications as someone who could and wanted to be helpful. Appearing in court with arrested members, hunting for jobs, helping with school problems, talking with parents, setting up ball games and money-raising events, holding "club" meetings, and, most important, simply accepting these kids as they were were the selling points. These, plus individual and group counseling, are the standard worker repertoire, and the Operators were already familiar with the routine.

So it all went well. Jacob pitched, they queried and tested, then he tested (their sincerity), and the contract seemed secure. Jacob would be the Operators' new gang worker, their "sponsor." The dozen table sitters were now quite animated, especially about club meetings, outings, and the status that came with being bad enough to deserve a sponsor.

But through the bantering came a lone calm voice from my left, from one of the quiet, breeze-coddled onlookers. He said it once but could be heard only enough for everyone to quiet down. He said it again, slowly and quietly but obviously as a direct challenge. "I—don't—think—so."

I looked at the door, gauging the distance I'd have to go before it all went down. What happened next is less clear in my mind because I stopped taking my mental notes for later recording. I was too fixed on the process. His name was Ranger, and Jacob asked him, "Why don't you think so?" The response was something like, "'Cause you didn't talk to us." The "us," it turned out, was Ranger, Pussyman, Yale, and the other three onlookers. And after perhaps five minutes, at the most, of low, threatening verbal exchanges, with Jacob bravely trying to keep the peace and also to assert his amenability to renegotiate, Ranger and his five filed out the door. The dozen followed, and all dispersed across the park grounds with shouts, laughter, slamming of park swings and slides, anything at hand.

Jacob hurriedly excused himself—"I gotta find Four-eyes"—and I took the first of many dark night walks from a gang meeting to my poor old, battered, carbon monoxide–filled car. In time, gang members learned it was OK to drive with me if all the windows were down. They thought a Doc ought to have a real car though, "Like a Lincoln, man."

A day later, I talked with Jacob and got the background. For two weeks he'd been hanging around the park connecting with gang members, one-on-one and in small cliques. "They're easy to spot," he said. "They're the ones hangin' around and watchin', but not partici-patin'" (Jacob always had trouble with those final g's). One day, sitting in the low bleachers watching a pickup football game, he spotted a fellow watcher—young, light skinned, athletically built, sweet faced—and so he moved over to talk with him, using the ball game as his opener. The young man was a local resident, temporarily unemployed but searching for a job. Jacob explained his interest in helping local youths and wondered whether this one had ever been in trouble. No, was the response, he'd never been incarcerated. Pleasant, quiet lad—not a target for Jacob.

But of course, this turned out to be Ranger, the one Operator who had more gang status than any other. "I missed him," Jacob confessed to me, "I flat out missed him. So I blew the meetin' by not knowin' who and what he was."

The end of my story is that Jacob mended his fences and became the Operators' worker—or "sponsor." And I got sucked into gang research, seduced by watching group process in the raw. For three more years, with a small staff of increasingly eager research assistants, we spent thousands of hours in contact with our gang members and their workers, about eight hundred of the former and a dozen of the latter. We learned much about gang member life which, with the occasional exception of a boisterous meeting, a fight, an exciting rumor, is a very dull life. For the most part, gang members do very little—sleep, get up late, hang around, brag a lot, eat again, drink, hang around some more. It's a boring life; the only thing that is equally boring is being a researcher watching gang members.

The project, using extensive data on the gang members and the workers, concluded that far from resocializing gang members and turning them away from criminal activity, gang workers had unwittingly increased gang cohesiveness, which in turn led to more crimes by gang members. This conclusion will prove to be central to many discussions in this book. The Los Angeles Police Department loved our findings: The gang "sponsors" were sponsoring crime. Their conclusion, too, will reappear in various guises in this book.

One of the responses to our findings came from the gang workers themselves: "Put up or shut up. If you can do better, let's see it." As a result, we sought and received a grant from a federal agency[8] to try out our ideas about cohesiveness and gang crime.

We applied the ideas to what was believed to be the most active street gang in Los Angeles; I'll call them the Latins. They were Chicano, structured identically to our earlier black gangs, a group of about the same size but with far more extensive criminal records.

There were three basic steps to the project, this time developed and managed and evaluated by the research team. If we were going to put up, then we were going to model it in our own image.[9]

The first step was to manipulate the environment and home in on gang members in such a way as to reduce gang cohesiveness. This meant cutting back and then eliminating "club" meetings and other group-qua-group activities; opening up alternative opportunities, such as the local Boys' Club, the teen post; getting tutors and big brothers and sisters; and, most important, working on employment problems. We failed to get much volunteer help for tutoring and big brothering, but we did quite well on the other issues. We chose our gang member targets on the basis of their relationship to the group—every choice was determined by our prediction of whether it would contribute to lowering the group's cohesion. We found more than one hundred jobs for them during the eighteen months of the project. That's not one hundred members but one hundred jobs; one guy went through ten and was eventually sent to prison for life (for murder, not for using up so many of our jobs). He reminds us that there are lots of losers out there, but the other forty-four who took and held jobs for a while are equally persuasive. Another member, the leader of a small clique, failed in his first two placements, showing off to his homeboys by arguing with his foremen. The third time we placed him by himself as an apprentice to a Chicano school-dropout who'd opened his own one-man print shop. Nando started to come home every night with printer's ink on his hands and arms. (If he washed it off, who would know he was a working man?) His girlfriend praised him. We praised him—publicly. He had money in his pockets. After a few weeks, Nando's two closest clique members couldn't stand it any longer. They asked us to get them jobs. We did, and all three became disappearing Latins.

The second step, measuring the changes in cohesiveness, was far easier, a technical issue of systematically gathering field observations on Nando's clique and all the others for eighteen months and subjecting them to statistical analyses. Depending on the measure used, we found a decrease in gang cohesiveness of up to 40 percent, not bad given that half the members were brothers or first cousins. Recruitment stopped completely after the first year. The cliques broke up into small segments, and to all intents and purposes, the Latins dissipated, the effect lasting for an additional six months of observations.

The third step, again, was comparatively easy, the measurement of delinquency, using police arrests for this purpose. Per-member arrest rates did not change substantially, but with the reductions in cohesiveness and recruitment, we found a 35 percent decrease in the number of offenses committed overall by the Latins. The connection between the reductions in cohesiveness and offense was quite high— dramatic in the first six months, with a continuing, moderate decline thereafter.

We were pleased, obviously. Our field-tested theory held up. Our convictions about group process and delinquency were reinforced. But to be honest, I wouldn't recommend that anyone repeat the process. This project was intensive, in both time and emotion. On various occasions we were at it day and night. We rented a room in the Latins' barrio and bought a used couch for $25 from Goodwill Industries to cover the late night and emergency situations as they arose.[10] Every move on every Latin was planned, programmed, monitored, and assessed. This was real work, without the evaluator's luxury of critiquing someone else's ideas.

Indeed, I was sufficiently worn down by the intensity that when a final gathering of a few Latins turned into a gang fight, I closed the book and declared I'd had it with gangs.[11] For ten years, until 1980, I talked less and less about gangs, and with greater and greater ignorance of what was happening. But then a sheriff's lieutenant with whom I'd worked closely on diversion evaluations told me that there was a brand-new gang worker program starting up in Los Angeles, maybe the biggest ever. The director wanted an independent evaluation of the program, "and I told him," said my friend, "that you were the right guy to do it. You have to. I know you said you'd had it with gangs, but you can't pass this up. You can't."

This became the next of the projects that shape this book, carried out, as have all the additional projects, in collaboration with my colleague Cheryl Maxson. It started with the request to evaluate the gang worker project. Cheryl joined the organization part time to help them develop their management information system, which would in turn provide data for the evaluation. But then the management changed: The gung-ho director upset the County Board of Supervisors and was replaced by a decidedly less dynamic bureaucrat, and that was the end of the evaluation. Nonetheless, we had received independent funding from the National Institute of Justice (NIJ) to assess the homicide rates reported by the LASD and LAPD, which served as the public measure of gang activity and presumably would determine the fate of the new gang worker program.

The new program came about in 1980 when Los Angeles County reported a dramatic new peak in gang-related homicides: The street gangs accounted for 351 deaths. Such an increase made many people suspicious about police-reported rates of gang murders: Were they deliberately overstated? Our analysis was carried out on the LAPD's and LASD's homicide files. Such files are very thick, ranging from a few inches to a few feet. They're filled with trivia. And sometimes, when you get to the pictures and the forensics, they can be stomach turning. We lost a research assistant who couldn't stand to extract data from another investigation of a drive-by shooting of an infant.

Our approach was to ask, statistically, what the contribution of three sets of variables was to gang and comparable nongang police investigations. One set described the homicide participants: age, gender, ethnicity, prior criminal record, relationship between offender and victim, and so on. The second set described the situation: number of participants, locations, presence and number of firearms or other weapons, time of day, auto involvement, and the like. The third set had to do with the investigative process: number of charges filed, witnesses' addresses located, witnesses interviewed, forensics, thickness of the package, and so forth.

In comparing gang and comparable nongang cases (from the same neighborhood, within a restricted age range of 10 to 30), we found that two sets of variables explained most of the differences between gangs and nongangs. Participant variables were the more important of the two, and situation variables were next.[12] The investigative variables contributed very little. That is, the way that the police investigated, recorded, and reported the cases had next to nothing to do with whether the case was gang or nongang. There was, in other words, little support for any claim of bias in the reporting of gang-related homicides. The homicide rates were in fact valid statements of gang-related deaths.

In addition, we found evidence that the police gang unit's involvement in gang homicide investigations resulted in better investigations. Police gang experts can offer intelligence on gang member suspects often not available to homicide detectives. This finding also brought us positive responses from law enforcement. It was a reminder of our initial conversation with the then Deputy Chief of Police Daryl Gates. We asked what would have happened if our evaluation of the earlier gang worker project hadn't supported the LAPD's position. "That's simple," Gates chuckled, "we wouldn't be having this meeting today." The chuckle was pleasant; the content was no joke.

OK, Los Angeles cops provided reliable and valid data on gang violence. But if this finding were unique to Los Angeles, then it was not of much value. The National Institute of Justice supported an additional grant to look at some of these same issues in five smaller, gang-involved communities elsewhere in California. We had to switch from homicide to a broader range of violence, since these communities had too few homicides to yield stable estimates. May they remain that way. What we found was quite encouraging. The same sorts of participant and situation variables that distinguished gang from nongang violence in Los Angeles also did so in these other five communities. It wasn't a perfect replication, certainly, but good enough to add confidence in the use of police statistics on gang crime as an indication of crime levels, not just of police activity. What we did not find in these five cities was confirmation that the existence or form of the police gang units made any difference in the clearance rates and filing rates in gang cases. That is, special gang expertise didn't seem to yield better investigations or prosecutions of cases. Perhaps that can't happen until gang units reach the level of sophistication exhibited by the gang units in Los Angeles.

We also became interested in the effect of different definitions of gang-related violence on police statistics concerning that violence. It's an interesting issue. The police in some cities such as Los Angeles define a gang incident by the presence of a gang member as either victim or offender. This member-based definition thus includes as gang crime a liquor store robbery committed by a gang member to fill his own pockets. In cities such as Chicago, a gang-related crime is defined by the police more narrowly by the involvement of a gang motive—retaliation, recruitment, turf battle, filling gang coffers, and so on. Accordingly, based on the motive-based definition, that liquor store robbery would not be tallied with the gang crimes. To be a gang crime, the robber would have to split the proceeds with his homies or would have to have selected that liquor store specifically because it was a hangout for a rival gang. Prosecutors, too, vary in their use of these two definitional approaches, with those in smaller jurisdictions showing a preference for the narrower definition.[13]

We analyzed hundreds and hundreds of LAPD and LASD homicide cases, using first the member definition and then the motive definition. Two findings emerged. First, the member definition used in Los Angeles yielded about twice the number of gang homicides. For instance, if the earlier 1980 peak of 351 gang-related murders had been measured instead by a Chicago-style motive definition, the peak might have been set closer to 175. It's understandable that Police Chief Daryl Gates pushed for the adoption of the motive definition, given the horrendous gang statistics coming out of Los Angeles.[14]

The second finding was, to us, equally interesting. Although the motive definition was thought to be the more "pure," the homicide cases it included were substantively no different from the member-defined cases that wouldn't be included by the motive approach. Both the participant and the situation variables looked much the same. One implication is that if you assign gang officers to participate in gang homicide investigations, you'll cut their caseload in half using the motive definition but lose their expertise in handling an equal number of similar cases. The balance between the public relations advantage of the motive definition and the investigative advantage of the member definition leaves the police in an interesting quandary. For us, as researchers, the member definition seems superior—twice the number of cases with no loss in validity. George Knox stated the case rather forcefully: "Cities that want to honestly confront the gang crime problem will use the Los Angeles definition; those that want to politicize the problem and obscure it hoping it will 'just go away' will use the Chicago definition."[15]

These analyses were completed on homicides recorded between 1979 and 1982. We've now completed a replication on homicides in 1988 and 1989 from both the LAPD and the LASD, this time with support from the Centers for Disease Control (CDC).[16] Gang and non-gang incidents differed in much the same way as they had earlier.

And although gang motive homicides had increased somewhat to 60 percent of the total, the member and motive definitions continued to yield vastly different numbers of "gang-related" homicides yet again shared many of the same characteristics of participants and settings. Thus the general tenor of the findings is stable, not a function of a particular moment in Los Angeles's gang history. This is an important point: The past ten years have seen more than a doubling of the reported gang homicide figures since what was once thought to be a one-time-only peak in 1980. Chicago, it turns out, is having the same problem. Its motive-defined peak in 1991 reached 129, according to the head of the Chicago Police Department's gang unit. This might be closer to 250 using the member definition.

If my 1960s projects in gang intervention constituted the first major block of exposure to gang issues, this more recent set of studies on the nature of gang violence and its reporting is the second block. It produced considerable confidence in understanding the extremes of gang behavior, that is, serious violence. This is important, because violence was simply not a major gang issue in the 1960s; it was there, all right, visibly so, but more a threat than an actuality. Gang violence in the 1980s and 1990s, however, is another matter.

The third block of studies has provided the opportunity to understand the most recent escalation of gang activity. In late 1984 there was a particularly bad incident in Los Angeles: Five people at a social gathering were shot and killed in what appeared to be a typical drive-by shooting, except that drive-bys are seldom so lethal. Ordinarily, one would record this in the gang retaliation column. But there was a suggestion—just a hint, then a rumor, then a claim—that the killings were a payback for a drug-related matter. The drug was something new, called "rock" cocaine. Rock cocaine, now generally known as "crack," has by now been around for sufficiently long that I don't need to dwell on it. But in the early 1980s, it was new and exploding fast, first in Los Angeles and, within two years after the shooting, across the nation. My colleagues and I were not drug experts and ordinarily would not have seen in the crack explosion yet another challenge. But what was being said about crack by the press and the police was right up our ally—that crack sales and distribution were in the hands of, in the control of, Los Angeles's street gangs, specifically the Crips and the Bloods. These were, and are, amalgamation names for a large number of both friendly and rival gangs. They were, and are, heavy contributors to the rising rates of gang violence in South Central Los Angeles.

The police and press version was that street gangs were ideal mechanisms for drug distribution (why only when crack appeared was never made clear). Gangs, it was said, were tightly organized, with a well-defined leadership, fierce internal loyalty, and well-established territories that they defended through violence. This gang image fit the needs of drug, or crack, distribution very neatly.

Except that this is not what most street gangs are like. I'll get into that more in Chapter 3, but for now it's enough to note that we saw gangs as relatively loosely organized, with an ephemeral and weak leadership and a loyalty rhetoric commonly suspended under stress or threat of prosecution. True, most gangs are territorial, but could this be translated from a sense of community turf to a drug business territory? True, gangs are more violent than other groups, but could that violence, generally expressive and often impulsive, be translated into controlled and planned drug business enforcement?

Everything we knew about street gangs said, "Hold on—that doesn't fit." We couldn't twist our understanding around enough to accept what we were hearing. Some of our police friends and contacts agreed with us—but not out loud. Local police management, including the LAPD, LASD, and other affected departments, were selling the gang–crack sales connection and selling it hard. Both evils were joined, so the police could throw up their hands, claim inadequate resources, and demand more help. Both Chief Gates and Sheriff Sherman Block ordered their own gang and narcotics units to work together, since this was now presumably just one problem.

Our response, as usual, was to line up the collaboration of the LAPD and the LASD and then submit a new proposal to the NIJ. The proposal was approved, and we analyzed all crack sales arrest incidents in the five most active areas from 1983 through 1985, the period of the crack explosion and stabilization. We also looked at the homicides during that period for any change in drug involvement. By that time, of course, we were getting quite efficient at homicide research.

What we found did not cement our relations with the police. At its highest point, we found evidence of gang involvement in crack sales arrests at just 25 percent; that is, three-fourths of crack incidents were not gang related. The evidence of violence did not equal that being suggested; in fact, we could find no evidence for an increase in crack sales violence attributable to gang involvement. To the contrary, when we looked at the homicide data, we found an increase in the nongang cases.

This was not what police (or press) had suggested, at least not publicly, and it was not what the NIJ anticipated. We sent our report on the findings to the LASD and received a polite letter from Sheriff Block acknowledging the likely correctness of the findings for that period but suggesting that gang involvement was still increasing; OK—reasonable response.

We sent the report to the LAPD—to the chief, the Police Commission, the head of the narcotics division, and to the head of the gang unit; no response.

We sent the report to the NIJ, which declined to publish it—out of date, I was told. Balderdash! The institute just didn't like the findings because they didn't fit with the institute's beliefs about gangs and crack. The NIJ rejected our research proposal to do a follow-up and update of the findings.

However, we've recently completed a small replication for 1988 and 1989 of the homicide–drug connection, and it confirms our earlier failure to find more violence in gang than nongang incidents. And a recent Centers for Disease Control study, using LAPD data for 1986 to 1988, matches ours point for point; out of date, indeed!

We also replicated the gang–drug study in two additional cities (funded under a different mechanism, under different management at the NIJ). The point is not to get yet more confirmation but, rather, to see how well the findings apply elsewhere. The findings from the project suggest, with some reasonable variations, that these two additional cities exhibit the same basic patterns as we found in Los Angeles. The gang contribution to drug sales is present, but not major. It's more a confluence of two independent forms of activity.

Finally, in this personal journey are two more studies constituting a fourth block of gang interests. Over the years, I'd been collecting a list of cities that reliably seemed to have street gangs. The list grew as I read new research reports, as our own research expanded outside Los Angeles, as police and other officials contacted me, and as media reports came in and could be verified. My younger daughter even chipped in with information about cities in Wisconsin.

At one point I had a list of more than 50 confirmed gang cities. By 1990, the list had grown to over 100. In just over one more year, I'd reached 187 and was still going. I started systematically calling and interviewing police gang experts across the country, including all cities with a population of 100,000 or more and a significant number smaller than that. Some cities were under 10,000 in size. By the end of 1991, more than 316 phone interviews had confirmed 261 gang cities. That's a horrendous number, and the search was not over yet.

In response to a solicitation from the National Institute of Justice, we received support to study the migration patterns of gang members throughout the country. This interest of the NIJ's stems directly from its continuing belief in the gang–crack connection and the persistent stories of crack franchises run by gangs nationwide, especially by the Bloods and Crips.

We sent a screening questionnaire to more than 1,100 police departments. The questions included most of those I used in my unfunded phone interviews. Out of these 1,100, including a random sample of smaller departments, we can now estimate the number of gang-involved cities and towns in America to be close to 800 and perhaps as high as 1,100.

This is the first time that these figures have been reported in print. I find them astounding, so much so that they demand amplification and explanation. I'll attempt both in Chapters 4 and 6. But keep in mind that the process is continuing. Between the date of this writing in late 1992 and the date of your reading, more towns have undoubtedly been added to the list.

Preview

The following chapters have a reasonably logical sequence. Chapter 2 sets forth in more detail the four issues or themes that run throughout the other topics to be covered. Thus Chapter 2 will provide more detailed coverage of definitional problems and offer more about the parameters of the proliferation of street gangs in hundreds of communities. I'll add some more grist to the gang–crack mill and from that move to a preview of the issues in gang process that are the building blocks of my positions on gang control. All four issues reappear in the remaining chapters, as they are tightly interconnected.

Chapter 3 gives the reader a historical overview of the gang situation in the United States. My purpose in this chapter is to be certain that the contemporary situation is understood in the context of change, of what used to be. All four issues of definition, proliferation, drug connection, and group process are dynamic; they take their present shape and significance in part from their departure from the simpler days, the "good old days," in which the original gang research was carried out.

Chapter 4 describes the street gang situation as it seems to be now. I say "seems to be" because the pace of change is so rapid and the research response is only now beginning to catch up. Chapter 5 covers many of the issues in gang intervention and control. But it's in Chapter 6, on gang suppression, more than any other that I need to drop the academic gloves and get more aggressive, if at some cost to collaborative relationships. The simple fact is that much of our local response and most of our state and federal responses to gang problems are way off base—conceptually misguided, poorly implemented, halfheartedly pursued. If I don't have all the answers—and that assuredly is true—I can at least challenge the answers we're used to hearing.

Chapter 7 looks at some developments that can be or are being considered. The battle isn't lost yet, but our armory is so depleted that we have to consider any new approach being suggested. Chapter 8 compares the U.S. gang situation with that in a number of other countries. Finally, I've added a brief Epilogue that permits me to return to a few points for emphasis and offer a warning or two; I guess that means another warning or two.

2

Basic Issues

Only thing when we first come up what we know is the streets. They give us a path where to go.

Former president, the Red Raiders

This second chapter lays out the four major concerns that weave their way throughout this book. They are, in order, (1) how street gangs might best be defined or described, (2) how widespread gangs have become in this country, (3) how the drug issue, especially the emergence of "crack" cocaine, relates to the current gang situation, and (4) the place of gang cohesiveness as a central, organizing concept for understanding the uniqueness of gangs and their control.

Definition and Description

The recent collection of gang articles edited by Ronald Huff includes three chapters that focus on the problem of gang definitions.[1] One calls for a new consensus; one shows how different definitions can lead to different conclusions about the seriousness of the gang violence problem; and one suggests that perhaps we should not seek consensus, for fear of overlooking important conceptual differences inherent in different approaches. Nothing could mirror our present state of uncertainty better than these three contrasting items.

It was in this context that I read one of the most dramatic and frightening depictions of criminal group behavior in memory, journalist Bill Buford's *Among the Thugs*. Buford's description of British soccer crowds—the "hoodlums," "the fucking thugs" as they call themselves—is situated in a conceptual array of crowds and mobs.[2] They are not street gangs. They come en masse to every game on a weekend, full of expectations of the excitement that is normative, anticipating their roles in the predicted violence, knowing just how to build up the tension and then deliberately propel themselves into their violent confrontations. Buford's description helps us exclude the soccer crowds as gangs, thereby further clarifying just exactly what a gang is.

Horowitz suggests that we shouldn't define *gang* lest we lose its essential variety. But too much variety in the concept defeats the chance to make any generalizations about gangs and at the same time leaves us talking right past one another, using the word yet attaching it to quite different referents.

If we can't find a universal definition—you'll remember how the captain derided my early attempt—we can try two other tacks.[3] First, let us exclude certain referents such as Buford's English soccer thugs. Second, let us review a few of the more commonly accepted attributes or parameters of a variety of gangs.

Since this book is about the American street gang, I'll add a third approach: Let us see what happens if we stress the street component.[4] History is important here. Thirty years ago, the commonly used phrase was *juvenile gang* or *delinquent gang* (with *delinquent* technically referring to offenses committed by juveniles). But as we moved closer to the current situation, gang membership slowly expanded upward in age. Our own early 1980s homicide and crack studies showed that the mean age of gang homicide assailants was just under 20 and that the victim's age was several years older. Our average crack-selling arrestee was also just under 20.

We and others, marking this age change, dropped the *juvenile* and *delinquent* terminology, and we heard the term *youth gang* more often. Spergel used that term for the title of his new book. But where does youth stop? Nowadays, a surprising number of gang members are in their mid- and late twenties, some in their thirties, and a few in their forties. They're hanging on longer and longer, and I'll discuss some of the reasons for this in Chapter 6.

As age becomes less of a defining characteristic, setting and style seem to add more. So now gang researchers are adopting the term *street gang*, which suggests something beyond age, something about the ordinary activities of gang members.

Exclusions

Our discussion, therefore, is of street gangs, not skinheads, motorcycle gangs, and other groups I have decided to exclude. There are younger gangs and older gangs, and some that span an age range as long as twenty-five years. But it's the street part of it I want you to consider for a moment. Most of the gangs that researchers write about—juvenile, delinquent, youth—are depicted as hanging around, usually in the open. It may be at a street corner, a taco stand, or on the side of a park watching the action; they're somewhere in the open or in the open behind a building. They're smoking, drinking, roughhousing, playing a pickup ball game, messing with a few girls, or sauntering up a street in a possessive, get-outta-our-way fashion.

Gilbert Geis wrote a sixty-page statement about gangs in 1965 for the President's Committee on Juvenile Delinquency and Youth Crime. The first three pages contain what is still the most charming statement in print on the etymology of the word *gang*. In it Geis anticipates my emphasis on the street component when he points to the early English usage of *gang* to depict "a going, a walking, or a journey."

Skinheads don't usually fit this picture. They're inside; they're working on their written materials; or if outside, they're looking for a target, not just lounging around.[5] Racist white groups in Bensonhurst, New York, recently given national publicity because of ethnic violence, were described as having "a hierarchy of desirable targets for assault"—blacks, then Dominicans, Pakistanis, and Indians. This yields "a sense of group cohesion and solidarity."[6]

Bikers don't fit this street gang picture. They're focused on their machines, cruising, or dealing drugs in an organized manner, and like the skinheads they're deliberately recognizable rather than recognizable by default.[7]

Street gangs seem aimless; skinheads and bikers are focused, always planning. Street gang members get into any and every kind of trouble. It's cafeteria-style crime—a little of this, a touch of that, two attempts at something else. Skins and bikers prefer narrower ranges of trouble.

By the same token, what I know of the heavy metal–influenced "stoner" groups, "punks," and the occasional satanic cult tells me that they don't fit easily into a category that includes street gangs.[8] Neither do terrorist gangs, who concentrate on narrow shared interests (cultural or political).

Writings on terrorist gangs are generally provided by journalists rather than by social scientists. One piece, by Luis Francis, describes the various levels of gang organization and violence in Manila, long known for its machete-hacking terrorist gangs. Another was provided by a Colombian student of mine who must remain anonymous. His contribution dealt with the *sicarios*, the young men on motorcycles who serve as hit men for the Medellín drug cartel. In one memorable instance, a *sicario*, long estranged from his family and on his own as a successful assassin, received a contract for a new hit. When the meeting was set up with the contractor, they turned out to be father and son. The father, one of the cartel's most notorious leaders, had no idea his young son had been doing the cartel's dirtiest business, so compartmentalized are these groups in their own crime categories.

Prison gangs have received a lot of attention over the past decade. Although media reports often suggest a close tie between prison and street gangs, experts generally minimize the connection.[9] Certainly it is true that many street gang members are recruited into the prison gangs when they arrive at the gates. Many feel they must seek out membership for personal protection. Adult prison gangs are generally founded along ethnic lines as protection against other ethnic groups. Thus the Mexican Mafia, Nuestra Familia, Black Guerrilla Family, and Aryan Brotherhood are notable for their ethnic homogeneity and interethnic gang fights. They also specialize in narcotics trafficking inside the prison, a line of work that often continues after their release (the film *American Me*, starring Edward James Olmos, provides a powerful depiction of this situation).

When street gang members leave prison, they generally return to their barrio or 'hood and reconnect with their street gang friends. In most cases, the lure of this group proves far stronger than that of the prison gangs. The strict rules and vicious sanctions that accompany the prison gang are not needed in the streets, and not wanted. Although prison gangsters may brutalize or kill their own members in the joint for code violations, on the street, homies don't kill one another.

Nor, on the less dramatic side, do I include youthful play groups or car clubs and "low riders," or street corner pals of the sort described in William Foote Whyte's *Street Corner Society* or Elliot Liebow's *Tally's Corner*. More difficult to exclude are the young "copy-cat" gangs or "wannabe" groups that are trying out gang life, or the so-called drug gangs that have become the subject of much discussion with the advent of the crack explosion. I'll return to both in a while.

In any case, this emphasis on the street gang permits a number of reasonable group exclusions. My later comments on the crime component will further delineate my core interest in this gang which, keep in mind, still comes in a dismaying variety of forms. But the narrowing of our focus will be helpful and has a good deal of consensual validity.[10]

The Criminal Component

Whatever may have been the history of the term *gang* and whatever may have been the desire—in many ways legitimate—to avoid stigmatizing youth groups with a pejorative term, it is time to characterize the street gang specifically for its involvement, attitudinal and/or behavioral, in delinquency and crime.

In this I am in some agreement with George Knox, who produced the first genuine textbook on gangs. He notes:

> Crime involvement of a group must not therefore be a sub rosa function, about which few of the members have knowledge, if we are to consider the group a gang. Members of many legitimate voluntary associations and civic groups are sometimes arrested for a variety of offenses.

But these are not offenses committed on behalf of their group; these are not offenses even necessarily known to their full social network; these are not offenses condoned and approved of in advance by their organization, or which enjoy their acceptance or blessing. To be considered a gang, the criminal involvement of members must be openly known and approved of as such.[11]

Let's take a practical example first. Asian gangs are included by almost everyone in the generic category of delinquent/youth/street gangs. Although there are significant variations among Chinese, Korean, Cambodian, Vietnamese, and other Asian gangs, they also share differences from typical black, Hispanic, and white gangs. They focus more on property crimes (and some specialize in the narcotics business), are better and more tightly organized with identifiable leaders, and are far more secretive. Most pertinent to this discussion, however, is that they tend to be far less territorial and therefore less openly visible. Their street presence is low compared with that of other street gangs.

Nonetheless, we generally include Asian gangs in the street gang category. The reason has to do with the balance of definitional criteria, the weight to be given to each. Asian gangs are somewhat lower on territoriality but higher on criminal focus.

In emphasizing the criminal component, some gang writers, like myself, have been attacked for "selling out" conceptually to law enforcement, that is, accepting the cop's view of the gang. That's nonsense; this complaint results not from our emphasis but from the contrary orientations of the critics. There really are two poles in this controversy. One defines gang without reference to gang crime, and the other emphasizes the crime element—usually violence—at the expense of all other elements. My position falls in between these two poles, but I do insist on including the crime component. This derives, in part, from the group-related perceptions of the gang members themselves.

Years ago, after my initial immersion in gang affairs, I became convinced that the very process of being accepted into a street gang led members to increased rates of delinquency. Our Los Angeles project data certainly suggested this conclusion through the correlations between group cohesiveness and commission of offenses. For instance, when comparing offenses often committed alone with those more often committed with co-offenders, it was the latter that were more affected by changes in cohesiveness.

But this was hardly proof of the importance of crime in the maintenance of group cohesion. Slowly but surely, however, other evidence has been accumulating.

In a small case study of one gang in Chicago, Leon Jansyn concluded that as the number of members began to decline beyond a certain point, those remaining deliberately increased their criminal activity in order to increase the cohesion among them. Paul Tracy compared nongang and gang arrestees in Philadelphia, showing far higher levels of crime among the gang members. Solomon Kobrin and his colleagues in Chicago found a positive relationship between a gang's status and its level of criminal activity.

More convincing evidence is now emerging from several longitudinal studies that allow one to follow the development of a delinquent career. Elliott and Menard used data from the National Youth Survey to demonstrate an interactive process in which beginning to hang around with delinquent associates is followed by involvement in minor offending, followed by association with more delinquent peers, leading to more serious delinquency, in turn leading to association with predominantly delinquent peers.[12]

It is not clear how many of these respondents to the National Youth Survey could fairly be listed as street gang members; rather, the value of the sequence described is to suggest the viability of the sequential process. The Denver Youth Project, in contrast, asked questions specifically designed to elicit admissions of gang membership.[13] Then gang members were compared with nongang offenders over several years. The data clearly demonstrate that even before joining, gang members were already somewhat more criminally involved than were nongang joiners; that joining gangs had the effect of significantly increasing criminal involvement; and that leaving the gang led back to a lower level of delinquency. Membership is selective for those with an initially delinquent orientation and then further enhances it.

In a yet more sophisticated analysis, Terry Thornberry and his group made similar analyses of gang and nongang data from the Rochester (New York) Youth Project, which confirmed most of the Denver results.[14] Denver is a more established gang city, whereas Rochester is quite new to the list, but the process is the same and helps confirm the impression derived from the classic studies of the 1950s and 1960s in Boston, Chicago, and Los Angeles. Joining gangs leads to more criminal involvement.

But criminal involvement in the gangs being studied still does not mean that it must be made part of the definition of a gang, so we're back to the two poles I set up earlier. Let's look at these.

Merry Morash published a study of "ganglike" youth groups in Boston. She used the term *ganglike* quite deliberately to emphasize defining gangs without reference to delinquency. She feared a tautology: If you define gangs as delinquent groups, then you lose the chance to investigate the members' delinquency independent of their gang affiliation. And as Morash defined her groups, and thus solicited her survey respondents, she might just as easily have been drawing from sports teams, boy scouts, or church groups. Look at her criteria:

> ...a regular place to meet outside the home; members typically belong for
> several years; group usually is found in the same place; group members

are drawn from a small neighborhood area; group has a name; core group is attached to a younger and/or older group; group frequently takes part in a variety of joint activities.[15]

For fear of the tautology, Morash has so broadened the definition of "ganglikeness" as to take us back to that enormous array of youth groups that invalidates any special aspect of the gang, a gang study that has an unknowable relationship to gang research.

A similar argument was made in several places by the admirable Jim Short, whose empirical work on street gangs in Chicago may still deserve the prize as "best yet." Short's concern is also with the tautological problem: If we define gangs in terms of delinquency, how can we then turn around and study gang delinquency as a function of gang involvement? We've talked about this with no resolution. His most recent words to me in a 1990 letter were as follows:

> I guess we simply disagree as to inclusion of delinquency in the definition of gangs. I certainly agree that group processes within the gang are critical to understanding why delinquency occurs when it does within and between gangs. Perhaps the most basic disagreement is in your use of "criminally inclined" to characterize gangs, since it appears to me that many gangs who commit delinquent acts are not so much criminally inclined as concerned about such matters as participation in youth culture, and getting by.

Others, on occasion, have also shied away from using criminal behavior to define gangs. A particularly lucid review and position statement was offered by Robert Bursik and Harold Grasmick, and I recommend it to the reader, despite their tentative conclusion: "We are uncomfortable with the delinquent behavior criteria, for it makes a possible outcome of gang activity one of the defining characteristics."[16] The flaws in arguments like these, it seems to me, are (1) how one can then avoid the kind of broad inclusiveness of characterizations like that of Morash and (2) the resulting tautology, which suggests an all-or-nothing statement of gang crime. But I have never suggested that gangs are criminal and that other groups are not. We all know better. It's not a matter of studying whether crime exists in the gang but how much is committed (and how it develops, how it differs, what effect it has on the members, and so on). Studying the level and nature of gang crime—as Morash and Short set out to do— is in no way precluded by finding that gangs are partly defined by their orientation to that crime.

Walter Miller has considered this issue for many years. Earlier, he was less convinced about the value of including crime in the definition of a gang. Morash's argument stemmed directly from that position. But later, in an outstanding essay on a number of gang issues, Miller noted,

One major area of disagreement concerns the role of criminal behavior in gang life. Although specifically illegal activities generally constitute a relatively small proportion of a gang's activities, they are often represented as its dominant preoccupation, or even as the basis of its existence. Thus, when youth groups in a particular community clearly appear to present a problem, they are perceived as gangs; when they do not, that community has "groups," not gangs. This essay treats illegal and problematic behavior as significant aspects of gang life, but does not conceive of youth gangs primarily as instruments for committing crime.[17]

There is great variation in the level and character of the offenses committed by gangs, and no definition has logically interfered with studying that variation. More important, I think, may be locating the "tipping point" in criminal orientation that helps us distinguish gang from nongang, at least as far as the street gang is concerned. I'll return to this point a little later.

First, let us look at the other side of the coin. This time, there are gradations, from those who want crime to be the essential definition of gangs ("It's a group of thugs; they're hoodlums, they're crooks and criminals"), to those whose definition is more specifically based on serious crimes, and finally to those who narrow the conception to violent crime in particular.

The essentiality is suggested by a piece written for the *FBI Law Enforcement Bulletin* by a professor-and-cop team. "A gang is a group of youths...banded together for antisocial and criminal activities," and street gangs "maintain the common objective of criminal activity."[18] In other words, youths form gangs and join gangs in order to commit crimes. It's a surprisingly widely held view and quite wrong in many, many instances. Crime and delinquency are more the product than the goal of gang joiners. There are more powerful incentives for joining—status, identity, a sense of belonging, protection—and I'll discuss these in the next chapter. But seeing crime as the source of ganging is merely a reflection of a narrower law enforcement exposure to gang members, producing an image then perpetuated by many media representatives and accepted by much of the general public.

The second variation is those who concentrate on the serious crime component. Here, I turn to my California setting for examples of overstressing serious crime. For example, the Los Angeles Police Department is now keeping its gang crime statistics according to two definitions, one requiring a gang motive in each incident and the other based on a gang member's being involved, regardless of motive. Only certain offenses are used for these statistics: homicide, attempted murder, felony assault, robbery, shooting into an inhabited dwelling, kidnap, rape, arson, witness intimidation, extortion, and—perhaps most symbolically—battery on a police officer. Inexplicably not included are all the thefts, burglaries, auto thefts, vandalisms, and myriad other offenses.

Now if we move up the enforcement ladder a bit to statistical reporting by the Los Angeles District Attorney's Office, we find a similar but somewhat expanded list: homicide, felony assault, *misdemeanor assault*, robbery, shooting into an inhabited dwelling, kidnap, rape (or *attempted*), arson, witness intimidation, extortion, *weapons law violations, narcotics crimes*, and—of course—battery on a police officer. I've put into italics the additions to the LAPD list, and as can be seen, the emphasis remains on the most serious offenses.

Further up the ladder, we find the list used by the California Department of Justice—in other words, the attorney general's list. This includes homicide, attempted murder, felony assault, robbery, rape, kidnap, shooting into an inhabited dwelling, and arson. This list is even narrower, a more serious listing than the previous two. Gangs, it all suggests, commit only horrific offenses and are thus altogether different from other offenders.

Finally, we come to some new and most interesting legislation that fully reifies "the criminal street gang" (section 186.22 of the California penal code), by defining it as

> any ongoing organization, association, or group of three or more persons, whether formal or informal, having as one of its primary activities the commission of one or more of the criminal acts enumerated in paragraphs (1) to (8), inclusive, of subdivision (e), which has a common name or common identifying sign or symbol, whose members individually or collectively engage in or have engaged in a pattern of criminal gang activity.

The criminal acts sound familiar: felony assault, robbery, homicide, narcotics offenses, shooting into an inhabited dwelling, arson, witness (or victim) intimidation, and vehicle theft. Those definitions relying on the most serious crimes portray gangs as not doing a lot of other things, which is patently ridiculous. This is legislation and regulation suited to limited enforcement and prosecution purposes; it has little to do with gang realities.

The legislation then goes on to describe gang membership in such a way as to permit increased sentences for gang-related crimes and for sheer membership itself if the member is aware of a pattern of criminal activity of others in the gang. This is really the culmination of defining a gang by way of crime, giving prosecutors and judges a greater capacity to incarcerate gang members for longer periods than they would for equally criminal nongang offenders. The legislation in question is known as the STEP Act—Street Terrorism Enforcement and Prevention—and will be discussed again in a later chapter when we look at law enforcement's gang suppression activities.

One last point on the criminal component: My insistence on its importance in defining the street gang is not directly related to the level of violent crime involved. Street gangs through the years have done nothing more often than they have done something exciting. Their most customary activities are sleeping, eating, and hanging around. Criminal acts are a minority of the activities they engage in, and violent acts are a minority of those. We must remember that despite the drama and lethality of gang violence, its prevalence does not deserve using the label *violent gang*. This only feeds a stereotype that needs no help from scholars.[19] To repeat, most gang members' behavior is not criminal, and most gang members' crimes are not violent. And of course, most violent people are not gang members, so its not very useful to define gangs in terms of violent crime alone.

Gang Parameters

Having excluded many groups from consideration and having stressed the street and criminal components of the definition (but without an emphasis on violence), you may have noted the absence of the more usual descriptors of gangs—age, gender, location, and so on. I turn to these now, rather than earlier, simply because these other gang parameters are far less controversial.

Briefly, gang members are young, usually between the ages of 12 and 30 and averaging probably around 20 years of age. They are primarily male; various estimates of female proportions range from zero to 30 percent. Most gangs—not all—are composed of homogeneous racial and ethnic minorities. Principally they are Hispanic or black, with an increase recently in Asian and other groups. They usually are territorial (*turf, barrio, 'hood*, and *set* are among the more common generic terms), although Asian gangs often present an exception. The gang's criminal activities generally are very versatile (the more focused pattern among "drug gangs" will be addressed later). Finally, street gangs range enormously in their duration, from a few months to decades of self-regeneration.

Even with the exclusions cited earlier, it may seem odd that a categorical term like *street gang* can encompass truly major variations in age, ethnicity, criminal pattern, and duration. It is no wonder that even with the exclusions, scholars and practitioners disagree on when a gang is "really" a gang; attaching the label inevitably becomes a bit arbitrary and a bit of an artful exercise. For instance, Jim Short views as street gangs the groups so effectively described by Mercer Sullivan. But Sullivan does not view them as such, nor do I.[20]

The "Tipping Point"

This definitional ambiguity has led me to a procedure for identifying street gangs in which I have developed considerable confidence. Play groups are not gangs, yet from the writings of Frederic Thrasher in 1927 to the present, researchers have described play groups that become gangs. The same is true of sports teams, break-dancers, "taggers," and other youthful collectivities. Where is the tipping point beyond which we say, "Aha—that sure sounds like a street gang to me"? I suggest two useful signposts.

The first, of course, is a commitment to a criminal orientation, a point that I have already stressed. Note carefully, however, that I specify orientation, not a pattern of serious criminal activity, as many in the enforcement world might require. There certainly are street gangs with a major commitment to delinquent and criminal activity. One can hardly be a gang writer in Los Angeles without acknowledging this. But as I have noted before, in even the most criminal gang, most of the members' time is taken up with noncriminal activities (especially if we include sleeping and lounging around as activities).

It is important, therefore, to emphasize the notion of orientation to crime, which also ties in to my second signpost, admittedly difficult to judge from outside the group: the group's self-recognition of its gang status. At some point, the gang-to-be starts to think of itself as a gang, a group set apart from others. My contention is that this self-recognition is based principally, although not solely, on the group's collective criminal or delinquent orientation. Very often, part of this process is an acceptance of intergroup, now intergang, rivalries and hostilities. It's hard to find a one-gang city; gang cohesiveness thrives on gang-to-gang hostilities.

Self-recognition is also fostered by a special vocabulary, clothing, signs and colors, and territorial graffiti, as represented in groups' names. As one member told me, "The only reason I look down on them was they were not Slausons. That was the only reason."

Finally, the gang's self-recognition of its character results from, and often reflects, the response it receives from segments of the community as the group becomes more troublesome. Schoolteachers and officials, social agency personnel, police, storekeepers, other nongang peers, many parents, and community residents have far less difficulty than scholars do in identifying a group as a gang. But here I am less concerned with their application of the label than with the group's acquiescence to it, acceptance of it, and eventually its stated pride in it. The community serves as a looking glass: Members look into it and see their gang character.[21]

In studying newly emergent gangs in St. Louis, Scott Decker sees the "tipping point" as revolving specifically and more narrowly around the orientation to violence.[22] This may be an overstatement for some street gangs, but it captures the same notion. I think Joan Moore has come to a similar view, that street gangs are groups that have gone beyond the "tipping point." As she concludes in her latest work: "In sum, gangs are no longer just at the rowdy end of the continuum of local adolescent groups—they are now really outside the continuum."[23]

The Proliferation of Gangs

The second basic issue to be addressed in this chapter probably has more to do than any other single factor does with the current public and official interest in street gangs. This is the fact that gangs have proliferated from a few to many hundreds of American cities. I say *proliferated* very deliberately, as opposed to *spread*, because the latter term suggests that the newer gang cities have been invaded by outside gangs. One often hears stories of Los Angeles or Chicago gangs' "spreading" to otherwise innocent urban centers.

But that's generally not the case. Most gang cities are facing indigenous street gangs—homegrowns, not imports.[24] There is much I need to cover about this proliferation of gangs because it is now such a visible component of America's burgeoning social problems. I'll devote a portion of Chapter 4 to this issue. Here, I want to describe the research process by which my colleagues and I have been establishing the prevalence of gang-involved cities.

I mentioned in the last chapter two surveys of police gang experts. The first evolved rather slowly as I became more aware of the extent of the problem nationally. It ended with personal phone interviews of 316 city police departments, of which 261 reported a genuine gang presence. Because the 316 departments were not a random sample, however, one should not conclude that almost 85 percent of U.S. cities have produced gangs. Rather, the survey was limited to all cities with a population of more than 100,000, plus a list of smaller ones that had come to my attention in a variety of ways.

Of the 189 U.S. cities that have a population of 100,000 or more, 176, or 94 percent, did acknowledge the presence of street gangs. Ninety-four percent is a striking level of gang involvement, enough to justify adding the questions I asked on a second survey of the prevalence of gang migration, funded by the National Institute of Justice. I'll report the results a bit later—they are most discouraging.

In addition, I selected a truly random sample of all cities with a population between 10,000 and 100,000. There are about 2,250 of these, so my sample of 60 is only a small proportion. Of these 60 receiving the written version of the questions, 23 of the 58 who responded, or 38.3 percent, reported having a gang problem. This is a horrendous number, since it suggests that perhaps 800 to 900 of 2,250 smaller cities in the United States are having problems with gangs, in addition to 94 percent of the 189 large cities. This amounts to about 1,100 cities.[25] The number could be an overestimate, as the questionnaire approach may have led to claims of gangs that could be discounted in an interview. Furthermore, one can take comfort in the fact that the average number of gangs claimed by police respondents who have such figures at hand in these 23 cities is only 5, or 3.5 if we omit the city of Compton, which reflects its close neighbor, South Central Los Angeles. The average number of gang members was put at 250, or "only" 113 if we exclude Compton. Thus, although a great number of police departments report that they now have a street gang problem, the vast majority are reporting a small gang problem.

The picture is dramatically illustrated in the data from the Maxson–Klein national gang migration survey, which included about 800 gang cities in the 1,100 queried about gang member migration known to the police. As displayed in Figures 2-1 and 2-2 on page 34, the proportion of cities with large numbers of gangs and gang members is very small. Most of the almost 800 cities able to report such data reported fewer than 6 gangs and fewer than 500 gang members. We have an increasingly widespread problem that is, nonetheless, not a large problem in most locations.

In contrast, Figures 2-3 and 2-4 on page 35 make it clear that larger cities have a larger problem. There's little surprise in this, and of course these are the cities we usually hear about in connection with street gangs: Los Angeles and Chicago are newsworthy, but Tyler, Texas, and Beloit, Wisconsin, are generally not.

Given these numbers, with their suggestion of the seriousness of the gang proliferation, I need to do two things at this point. First I will summarize the results of the 261 phone interviews much as I reported them to the more than three hundred police chiefs and the gang experts who responded to my interview questions. Then, because the reader should be skeptical about my belief that these police respondents were being reasonably honest and accurate in their assertions about local gang situations, I will provide examples of typical interview dialogues to give some sense of how they went.

Interview Summary, 261 Notable Gang Cities

Most of the respondents for the cities I surveyed were gang investigators or gang unit supervisors, although some were higher in the police hierarchy. Their expertise in and direct knowledge of street gang affairs were generally quite high, leading to data of far greater validity than have generally been available to and used by the media or the federal and state officials from whom the public has generally learned of gang matters. The following items summarize what I was told by these gang experts.

Gang Emergence. Forty-one of the cities experienced street gang problems before 1965. Another 85 first saw gangs emerge between 1965 and 1984, and 134 newly experienced them from 1985 on. It's clear that the problem has been escalating rapidly. Almost none of the cities reported having had a gang problem earlier but not now; the problem rarely goes away.

Migration. Much publicity has been given to gang problems in smaller cities being caused by Los Angeles and Chicago gangs spreading elsewhere. The gang experts in these cities did not paint such a picture. Although many of them noted some in-migration of big city gang members, relatively few felt that their gang situation was actually initiated from outside. Most gang problems, it seems, are locally derived, although in a number of cases, they are exacerbated by traveling gang members from outside.

Migration sources are widespread, with ninety-three different cities mentioned. However, the Los Angeles area, with 27 percent, and the Chicago area, with 18 percent, together account for almost half the mentions of migration sources. Much of the gang member migration from each of these places is to other cities in the same areas, whereas much of the remainder is also local or regional. Thus the migration of gang members yields a mixed picture. Heavy migration exists but is unusual; light levels of migration (a few outsiders) are more commonly reported, but most common is the failure to report any migration at all. And again, contrary to many media and other reports, the bulk of gang migration seems to be regional or local rather than national. When they knew the reasons for migration, these police respondents cited normal residential moves by the families of gang members as more common than moves deliberately designed to reap the benefits of the drug trade.

Size of the Problem. A number of these cities reported large numbers of street gangs and gang members, but far more are facing a smaller number. Ninety-seven cities reported 5 or fewer gangs; 46 cities reported 6 to 10 gangs currently active; 36 more cities reported 11 to 20 gangs; 19 reported 21 to 30 gangs; and 15 more reported 31 to 40 gangs. The numbers then dropped off even more rapidly. In sum, almost 60 percent of all the cities reported 10 or fewer gangs. In contrast, Los Angeles City and County together count up to 1,000 or more street gangs.

These gang numbers must be viewed cautiously, of course. Different departments use different definitions of a gang; some count only "core" members or "active" members; and some gave me very rough estimates because they had not carefully recorded the numbers. Typically, cities reported the total number of gang members to be in the hundreds; indeed, 63 percent reported fewer than five hundred gang members, and only one-fifth reported numbers in the low thousands. The general pattern seems clear: In the majority of these gang-involved cities, the numbers of street gangs and members are enough to call forth official recognition, but not so large—as yet—to constitute a truly major enforcement problem.

But also there are notable exceptions to this description, made worse in some instances by the relationship to drug markets and violence levels.

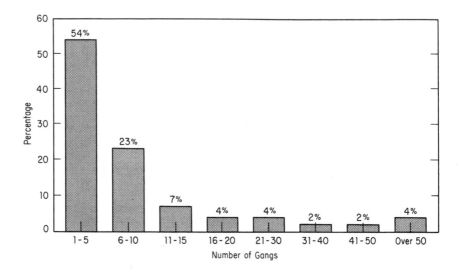

Figure 2-1. Numbers of Gangs Reported by 766 Gang Cities.

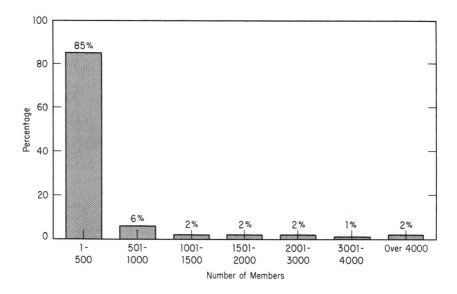

Figure 2-2. Numbers of Gang Members Reported in 739 Cities.

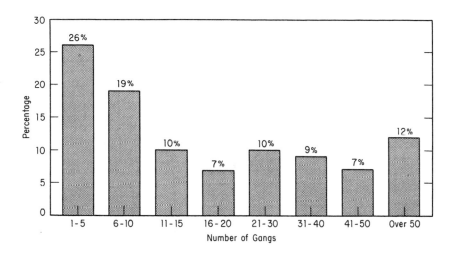

Figure 2-3. Numbers of Gangs Reported in 175 of the Largest U.S. Cities (>100,000).

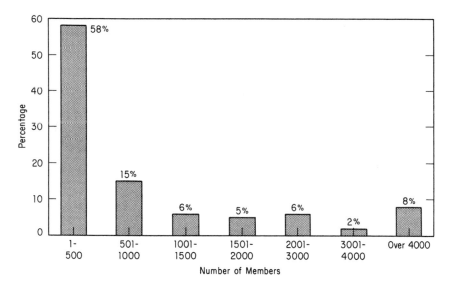

Figure 2-4. Numbers of Gang Members in 173 of the Largest U.S. Cities.

Structure. The old, traditional gang structure of past decades seems to be declining. The street gang with up to several hundred members, broken down into age cliques of "midgets," juniors, seniors, and *veteranos*, or "original gangsters," is giving way to relatively autonomous, smaller, independent groups, poorly organized and less territorial than used to be the case. In many cases, these police experts now have a hard time keeping up with the ever-changing ebb and flow of these smaller groups.

Ethnic changes were also noted. Although black and Hispanic gangs still predominate, many cities are now reporting Asian and South Pacific groups, small numbers of white (often supremacist) groups, and more ethnic mixing than used to be the case in gangs.

All of this, I sensed, is making police intelligence on gangs more difficult, with more need to share experiences among departments.

Drugs. In past decades, drug use was common among gang members, but organized drug distribution was not. This situation has changed somewhat with the crack explosion, but according to these police interviews, this change has been exaggerated by the media and many public officials. Only 14 percent reported a strong connection between gangs and crack, in which street gang members played a major part in drug distribution. Separate "drug gangs" or "crack gangs" were reported in 16 percent of the cities, but in most cases these did not comprise the majority of the gangs. Others commented on crack or other drugs as exacerbating an already developing street gang problem, and some respondents took some pains to emphasize that in their cities, crack was not the problem, that they had no special drug gangs, and that the reports of Los Angeles, Chicago, or Miami franchising of crack markets was not a part of their experience. Indeed, 72 percent reported the gang–crack connection to be moderate, weak, or nonexistent. I was being given a very deliberate message by these gang experts, who seemed to be reacting to a media image they found quite unjustified.

Departmental Responses. Police departments have varied widely in their response to the gang problem. Initially, for several years, over 40 percent denied having a problem, whereas others quickly responded with a call for action. Eventually, even the deniers had to take action (often because of a particularly horrific incident or the victimization of a public official's family member).

Typically, a small unit was established (one person or several) to gather gang intelligence, although in some cases a larger task force was organized to attack the problem. In some instances, responsibility was lodged with the Detective Bureau, and in others, Patrol or Narcotics was given the job. Most departments in seriously gang-involved cities now have a gang unit of some sort.

The intelligence function is almost universal in these units. A far smaller number participate in other functions, either prevention and community education or investigation of gang-related crimes. About fifty-five (or 20 percent) become engaged in occasional gang suppression activities, the most common of which are street sweeps, saturation patrols, hot-spot targeting, and general "harassment" activities. One department, at the extreme, developed a secretive, black-robed "ninja-type" attack team, uncomfortably reminding critics of the terror squads in some police states. At the other extreme, some of our respondents reported with considerable feeling that they would object strongly to suppression-style functions. If intelligence gathering is the common denominator of all gang units, it is clear that there is little consensus on what else they should be doing.

In sum, my police respondents gave some very candid and thoughtful interviews, including their view that what they don't know is often more than what they do know. Many seemed relieved at my willingness to accept their professional estimates in lieu of documented reports. This much is clear: The gang problem is exploding; it cannot simply be explained by the drug connection (or, therefore, solved by controlling drugs); given the current numbers and the continuing decline of inner-city life, there's plenty of room for the problem to increase; law enforcement has not yet achieved consensus on how to attack the problem.

The Police Gang Interviews as a Valid Source of Data

Some readers may not be as satisfied as I am with the quality of the information provided by these interviews. The reason is that no one else heard the respondents and no one else was available to judge the care and caution with which the great majority of them entered into the interview conversation. The fact that my colleague Cheryl Maxson kindly took on a few of the interviews and had similar results and reactions to the process is a comfort to me, but probably not to the really skeptical reader.

Let me paraphrase a few typical interview protocols to give a sense of how they went. Since none of them were electronically recorded, what follows are "typical" interview components taken from my notes, after the introductions were completed.

Respondent 1

Q: Leaving aside bikers and stoners and prison gangs, so we're concentrating now on street gangs, when would you say they first became apparent in your jurisdiction?

A: Well, of course, that depends on what you mean by street gangs. What's your definition?

Q: I guess I'm really more interested in what your department means by street gang. You said you have some, so what kinds of groups would you include?

[The respondents then generally described the targets of their interest, occasionally answering my questions about structure, ethnicity, crime involvement, and so on.]

Respondent 2

Q: You said these groups are "wannabes," kids in school who are showing gang colors, hand signals, and the like. Let me ask a couple of things: Do they have rival groups that they fight with?

A: Well, not that I've seen—lots of talk, but it's just bravado.

Q: OK—what kind of problem do they pose for you—are they committing much crime?

A: Well, no—it's mostly petty stuff that we can handle in the normal way. They think they're tough, but so far, like I said, it's more talk than action. None of your Los Angeles stuff yet.

Q: Then do you feel comfortable calling them gangs, or do you think they're not there yet—do I put your city on the gang list or the potential list?

A: Naw, we're not really into gangs here, not yet. But we got our eyes wide open. We know it's coming!

Respondent 3

Q: When would you say they first became apparent there?

A: In 1988, we grabbed some Crips coming in at the airport with the idea of starting a crack market here.

Q: So your first contact was with outside gang members? Did you have some homegrowns of your own?

A: Not at first. But when the crack came in, they tried to organize to sell the stuff. We cracked some heads, and they never got very far.

Q: So at this point you don't really have a gang situation?

A: No, not really—it's just that they keep coming back to it, trying to get a foothold.

Respondent 4

Q: I agree, it depends on what we mean by gang. Can you describe the kinds of groups you've got?

A: Well, we got two types. The Hispanics are real territorial, and most of their violence is against each other. They got midgets and hardcores and veterans, you know—a whole traditional way of doing things.

Q: OK, they sound fairly typical. What about the others?

A: Asians—we got Vietnamese mostly, and some Cambodians.

Q: Are they ganging?

A: Yeah, but not like the Hispanics. They don't have territorial notions. They're more into extortion stuff against their own people, and a lot of "house invasions" because these people don't trust banks and keep their money at home. The gangs bust in, very brazen, and threaten all kinds of stuff if the folks don't fork over the money. It can get real nasty at times.

Q: But is there real gang action, as you see it, or just a lot of free-lance racketeering and robberies?

A: Oh no, it's real organized and patterned. These gangs have their own names and leaders, and they communicate with each other from city to city. It's just that they have a real material purpose, not just defending themselves like the Hispanics. They're crime groups of young guys. The streets are just one of the hangouts.

I should point out that not all interviews went so well. Some examples of my notes are "Terrible interview"; "This guy is not gang knowledgeable—too far from the scene"; "He's harping back to his own gang involvement, not reflecting the present"; and "Bull-shit artist—clearly not a reliable informant".

But the point about these interviews is that most of the respondents were, by and large, being thoughtful. They quickly became involved in the definitional issue and understood its importance. They became research informants as well as mere respondents. And those that were new to the work or really were not familiar with gangs quickly revealed themselves. There was one who listed the Black Panthers as his town's first street gang—a stereotype cop response, not a thoughtful one. A few other referred to their own childhood experience as gang members and could only generalize from that. Their reports could be discounted fairly easily.

I used follow-up probes on gang structure, ethnic differences, size, and age distribution. If such questions yielded vague guesses, great care was taken before accepting the city as a gang city; I was trying for confirmation, not the largest number of gang cities I could get. Finally, I learned much from questions about the departments' responses to possible gang problems. Was there a denial phase? What led to the change from denial to acknowledgment? Is there a gang unit? If not, is there someone specifically assigned to gang intelligence? Are these groups dealt with any differently than other criminal suspects? Is there some agency or "civilian" in town who is knowledgeable about the gang situation?

In the end, I think there were few respondents who could claim the existence of gangs where there was none and few who could successfully cover up a gang situation. A few of the latter—Tulsa, Jacksonville, Pittsburgh, Tampa—switched in midstream. When I get to the large number of gang cities in Chapter 4, it is wise to remember that those data are a combination from two sources: the personal interviews in 261 cases and the questionnaires returned as part of a larger gang migration study. I have less faith in the questionnaire data because there was no chance for the kinds of dialogue described earlier. Still, the same basic questions were asked, and in many cases we have interviews and questionnaire returns from the same departments, yielding good estimates of reliability. The questionnaire data should not be dismissed easily—the gang situation, however minor it is in many cities, is truly widespread. This fact must be acknowledged and policy implications considered seriously.

The Connection Between Gangs and Drugs

More than any other gang issue—more even than gang proliferation or reported levels of gang violence—it is the place of gangs in drug sales and distribution, especially of crack cocaine, that has epitomized attention to gangs since the mid-1980s. I'm convinced that the connection has been blown way out of proportion by some zealous police officials, federal enforcement leaders, issue-hungry politicians, and headline-hunting media reporters. Nonetheless, their assessment of the connection has made it a major issue, and I can't ignore it here.

An example of just how committed to the gang–crack connection some enforcement officials have become is provided by the deputy chief of one large department. His duties included overseeing narcotics operations in his city, which also had a serious gang problem. "Look," he told me, "this narcotics stuff is all a matter of gangs and conspiracy. To me, a gang is any two or more guys working on crime together. In a drug sale, you got at least the seller and the distributor involved. Now that means it's a conspiracy. And there's two guys, right. So all these crack sales are gang crimes." [I asked for clarification, elaboration, help in being sure I understood him correctly.] "Damn right," was the answer. "Two or more guys conspiring to make crack sales means it's a gang affair...that's how we define gang around here."[26] The logic here is quite astounding, twisted, and turned to guarantee by definition an intimate connection between gangs and drug sales.

I want to repeat what I suggested earlier and will elaborate in Chapter 4: that half the connection is pure hype; the other half exists and is worthy of attention along with other equally important gang issues, such as proliferation, crime patterns, ethnic variations, and approaches to control. The reality of the drug sales connection does not match its press. The reason that gangs don't normally control drug sales (mainly crack sales) is also important for what it teaches us about the essential nature of street gangs, their structures, and their cohesiveness.

After lecturing on the drug–gang issue in some very public settings, my comments were picked up by the Associated Press and amplified following a rather long interview by an AP reporter. I soon received copies of the AP story from a variety of sources around the country and phone calls from colleagues who were surprised to see their media-avoiding compatriot "coming out," as it were.

The most interesting call, however, came from an official of the federal Drug Enforcement Administration (DEA) in the Midwest. It seems that my comments were the topic of a supervisors' meeting, and this caller had taken on the task of straightening me out. The conversation was a bit unpleasant at the beginning but became far less so as we chatted.

His pitch was basically that—of course—I didn't know what I was talking about because the DEA experience was that crack distribution was an organized affair, a conspiracy, and therefore in the hands of gangs. I asked the following:

- Was the conspiracy essentially at the mid-level and cross-border levels of the action? Answer: "Yes sir." I reminded him that my reports pertained primarily to lower street-level dealing, where the DEA is not normally involved.
- Were the "gangs" with which he was familiar local street gangs, or were they mostly focused on drug selling, on marketing a product, rather than defending turfs, engaging in a wide variety of non-sales-related crime, and so on? Answer: "They're drug gangs—all their criminal activity is driven by the money in drugs." I explained that street gangs had been the topic of my remarks and my research and that drug gangs, as he saw them, were not normally street gangs and certainly did not comprise the majority of gangs.
- Was he talking mostly about black gangs selling crack, or was he seeing other ethnic groups heavily involved? Answer: "No, sir, they're mostly black—a few Jamaican and other groups, but it's the blacks." I reported to him that Hispanic gangs are a major component of the street gang situation, that some Asian and white gangs are also out there, and that my remarks referred to all street gangs, not just the black gangs, some of whom were involved in crack sales.

The conversation continued at some length as we clarified the distinctions between street gangs and the sorts of groups the DEA was being thrown against. *Drug gangs* and *street gangs* are by no means synonymous. The national hype about crack-selling, crack-controlling street gangs derives in part from the unconsidered assumptions (and, in some instances, the deliberate assertion) that drug and street gangs are one and the same thing. It just ain't so!

It is useful to recall the data I reported a bit earlier from my 261 police interviews. Only 14 percent reported a strong gang–crack connection (86 percent did not). Only 16 percent reported having drug gangs among the total complement of gangs in their cities (84 percent did not). Moreover, in those cities, black gangs were slightly in the minority. Since it is black gangs who are principally involved in crack sales, then already there's a built-in ethnic "cap" on the drugs–gang connection. Both data and logic argue against the kind of involvement or control of crack still claimed by many federal and state officials and a few academics whose data are limited by the cities in which their work was done (see Taylor on Detroit, Padilla on Chicago). There is yet another logic that also calls a major gang–crack connection into question, and that logic derives from the very nature of street gangs. I'll get into this more a little later, but let me state it succinctly here. A drug-sales conspiracy, carried out by any group of people, requires the following for its success:

1. A clear, hierarchical leadership.
2. Strong group cohesiveness.
3. A code of loyalty and secrecy.
4. A narrow focusing of efforts on the mechanics of drug sales and the avoidance of independent or non-sales-related criminal involvement.

None of these four features is characteristic of the typical street gang. Forget your *West Side Story* image of gangs; instead, they tend to have shifting leadership, intermediate levels of cohesiveness, frequently broken codes of honor, and very versatile and independent criminal involvements. Gangs are lousy mechanisms for drug distribution.

I met a social agency director from Camden, New Jersey, who had extensive contacts with drug gang members in her community. She was absolutely dumbfounded, disbelieving, when I went through the preceding little litany about street gangs. All she knew was drug gangs, and so she assumed all gangs were like that. After two days of interaction, I got her to understand that the Camden gangs were not a mirror of the American street gang, but it did take two days. It's hard for people immersed in the drug business, be they DEA agents or Camden psychologists or Detroit researchers, to realize that their world is a restricted one. After all, before 1984 there was no crack in our cities, yet there were thousands of gangs. Crack is just an add-on, and only in some locations.

Gangs—street gangs—aren't well enough structured to suit the sellers' purpose. John Hagedorn describes a gang in Milwaukee that tried out the entrepreneurial drug-selling style. "Why did they stop when they were making money?" asks Hagedorn and answers,

> It was too much of a hassle. This gang was not cut out to become an organized criminal venture. The police did not really bother their sales,

but the organization necessary to pull off an on-going drug house was too much for this gang, whose members were concerned with "making it" day by day. While many of the adult members of this gang still sell cocaine and marijuana, it is done, as in most other gangs, individually and sporadically.[27]

Scott Decker and Barrik Van Winkle studied these issues in another midwestern city recently beset by gang problems, St. Louis. They report that the traditional gang structure of the 1950 and 1960s (see Chapter 3) has emerged in St. Louis—they call it a gang "constellation"—and it is too disorganized to support effective drug distributions. Members sell; gangs don't. Profits are used individually, not for the group. Drug sales are very low in the hierarchy of reasons for joining the St. Louis gangs.[28]

The major points to be remembered about gangs and crack—gangs and all drugs for that matter—are simply these: (1) Use is far more common than sales; (2) sales involvement on the whole is rather low; and (3) there is enormous diversity between and within cities in the level of gang involvement in drug sales. Consider the following contrasts: Crack sales involving gang members are at moderate levels in South Los Angeles but are almost nonexistent in the worse gang areas of East Los Angeles. Drug sales have long been connected to street gangs in Chicago, yet crack has yet to become prominent in that city. Crack sales are phenomenally high in the nation's capital, yet Washington has never been a gang-involved city.

Obviously this list could be augmented—my survey did some of this—but the specifics are less important than the pattern of diversity. This diversity tells us that there is no necessary connection between crack or other drug sales and street gang activity. The connection is city specific, ethnicity specific, and neighborhood specific. The justification for broad national or state policy based on the gang–drug connection is all but nil at this point. Only ideology requires such policy.

The Centrality of Gang Cohesiveness

I suggested earlier that gang cohesiveness is central to the nature and control of street gangs. This is the fourth of the basic issues weaving their way through this book. There are several reasons for emphasizing cohesiveness even more than gang leadership, crime patterns, gang etiology, and so forth. First, cohesiveness is the quintessential group process: Our understanding many aspects of gang leadership, crime, etiology and so on must derive in part from group process. The gang is far more than just a collection of individual persons. Second, cohesiveness gives clues to gang crime patterns—it accelerates violence; it interferes with drug distribution; and it accentuates some crime targets over others. Third, the cohesiveness of gangs is pivotal to the gangs' response to efforts at intervention and control, as I shall note more thoroughly in Chapters 5 and 6.

In the 1960s, the Ford Foundation provided funds to a special gang unit in the Los Angeles Probation Department—the Group Guidance Section, it was called—to demonstrate its ability to intervene in a growing gang problem and turn it around. The Group Guidance operation had been in existence since the late 1940s, working principally with traditional Hispanic gangs. It was determined, late in the 1950s, that black gangs were emerging in South Central Los Angeles and threatening to become an even more serious problem. Thus the Ford funds established a special emphasis unit within Group Guidance to connect with (*sponsor* in the gang members' terms) four large black gangs. The project was initiated in early 1961 and had four years to show what it could accomplish with one female and four male gang workers, county-paid student workers, and a unit supervisor.

Ford also funded an ongoing evaluation of the project, and that's where I came in. The evaluation, carried out at the University of Southern California, was foundering with a halftime director located across town at the University of California at Los Angeles, many miles from the small research staff and the project area. A major change was needed, and I was it. Significantly, I was seduced not only by the intriguing gang processes I observed but also by the dedicated efforts of the gang workers to make a difference.

Their basic routines were three: to carry out group activities with their assigned gangs, such as weekly club meetings, sports activities, money-raising events (car washes, dances, etc.), and skills development, most notably a tutoring project; to carry out individual counseling on family, school, and interpersonal (and sexual) relations and on delinquent behavior; and to open up the institutional doors in schools, court, job opportunities, welfare agencies, and the like in order to "de-isolate," as they put it, the gang members from their own community institutions.[29]

Pertinent to the discussion of cohesiveness were the difficulties the workers were having in getting support for more aggressive programming, which included lukewarm support, at best, from their own department. The workers were having trouble finding meeting places, finding volunteer tutors, finding employment consultants, and finding supportive school, welfare, and court personnel. And the police were actively antagonistic. Yet here were five street workers in direct contact for several years with almost eight hundred of the most agency-resistant, delinquently oriented youth in the city. The workers were accepted by the gang members; they were respected by them; they were listened to; and the gang members seemed to respond.

So I did what I could to help augment the program, to open those doors, to argue for more support from the Probation Department, and to get more group activity going with these gangs. I think I was, in fact, of some help, and, as I discovered later, to that extent I helped make these gangs more delinquent, more resistant, and more isolated from their own communities. That, at least, was the overall effect of the project, and I had become an advocate of the project.

How did this happen? By increasing the gang's cohesiveness is the answer. I'll share with you some of the more demonstrable items that support such a conclusion.

1. In the very midst of the project, in the spring of 1963, it was suspended for several months because of a bitter controversy with the Los Angeles Police Department, which claimed that "sponsored" gangs created four and a half times as much crime as those not getting Group Guidance attention did. The project responded with the fire analogy: You send the firemen where the worst fires have broken out. That is, the gangs being dealt with were deliberately chosen because of their unusual levels of criminal involvement. Both sides used the same data to draw opposite conclusions. Our own data offered little support, at that early point, for either conclusion, but then, neither side thought to ask us, the project evaluators, if we had anything to contribute, anyway. Ideology does not respect independent assessments.

In any case, when the project started up again after a *pro forma* agreement between police and probation—an agreement that changed not one iota of the project's goals or practices—the two most aggressively antipolice, progang workers asked to be transferred elsewhere, and they were. From that point on, those two gangs disintegrated. Newly assigned workers could not bring them together, could not arrange club meetings, and could not even find their gang members on the street. "What's going on, Doc?" one of the new workers asked me. "I got a roster of 125 names, and I can't find these dudes anywhere." I suggested he try dropping by their homes, a suggestion he found quaint—the "Klein theory," he called it—yet that's where he found the bulk of them, at home.

What had happened, I believe, was that the original two workers had inadvertently become the focus of the gangs' cohesion. Their active group programming, their antipolice attitudes, their total commitment to their groups had become even stronger glue than the members' original need to come together for identity, status, and belonging. Indeed, one worker remarked, "It'll take a new worker a year to get where I am with these kids." If true, this was a statement of failure, not success. His supervisor liked to point out, quite correctly, that a gang worker was successful if he worked his way out of being needed.

Throughout the period of this project, the research team gathered data on gang cohesiveness from field observations and the detailed logs required of the workers. Following the suspension of the project, we charted the cohesiveness of both these gangs going steadily downward, whereas it had been increasing before the suspension.[30] But the same data for the other two gangs, those that retained their workers, continued to show an increase in group cohesiveness. Although the decrease in the worker-transferred gangs is suggestive, the contrast with the worker-retained gangs is more convincing. Active gang workers, by the very nature of their activities, can almost not help but tighten even further the groups to which they are assigned.

2. Interesting examples were noted in all four gangs involved in the worker-induced recruitment of younger gang members (who would, in today's terminology, be called *wannabes*—youngsters who want to be accepted by the gangs). When a few younger boys (and girls, on occasion) were noted hanging around the "club" meetings, each worker responded in basically the same way. Since they were hanging around, they must need attention; if they needed attention, the worker was the self-appointed attender. He would acknowledge them, talk to them, clarify his role, and eventually help them form the nucleus of the next generation. They became the "Unborn Vikings," the "Baby Gladiators," the "Lil' Jokers," or the "Baby Bartenders." Perhaps they would have formed up without the worker—traditional gangs by definition regenerate themselves from below. But it was clear that the workers' presence and mode of operation either created or accelerated the new subgroups. In the pejorative LAPD parlance, they "sponsored new gangs."

3. As we noted more and more instances of cohesiveness building through actual gang worker intervention, we recognized the epitome of the process in one worker and his group. Manuel had replaced Jacob in 1962 (I'll describe Jacob's approach to his new gang shortly). An ex–gang member himself, Manuel quickly became the most active group activities initiator of all. He had "club" meetings once and sometimes twice a week. He piled kids into his truck—packed well beyond legal limits in many cases—and transported them thither and yon for various sports activities and other events designed to reduce their social isolation. He constantly used the group as his field structure, his way of filling hours away from the office.

Our observational data and worker reports soon reflected the effect: More members appeared, more participated, more joined, more increased their "street time." Here was a corroboration of our other observations and also an opportunity to test their value. We discussed the situation with Manuel's supervisor, the one who said that the worker's job was to work himself out of his job, and it was agreed to cease all of Manuel's group activities: no more club meetings, no more outings, no more informal gatherings at the park. From now on, Manuel was to be a one-on-one worker, a home visitor, a "sponsor" of individual gang members only.

Manuel, understandably, kicked and screamed. He was a group worker; we were taking away his crutch in the ambiguity and loneliness of fieldwork. But with an active field supervisor and the omnipresent research observers watching, he had little choice but to comply. The results were rather immediate and very dramatic. Our measures of cohesiveness showed close to a 50 percent drop: The gang members reduced their appearances together on the street, stopped hanging around the park as much, and—as far as we and Manuel could discern—did not invent new settings around which to coalesce. There had been one store where members often gathered, and this continued following the cessation of planned group activities. Manuel stayed away from it on his own, explaining, "That's their special place; I don't feel right going in there." From our point of view, we were pleased that he didn't. The decrease in cohesiveness was our test, and Manuel's gang passed it.

4. I mentioned Jacob's transfer to another gang, having been burned initially in his aggressive approach to the Operators. After his transfer, again to replace another worker, he decided to play it cool. He sat, very visibly, behind a large agency window fronting the main arterial walkway of his new group. Soon a handful of members approached him about starting up "the club" again. He declined—"They pay me just for sittin' here watchin' you guys. I'm not spendin' my nights with you, too" (club meetings were held generally in the evenings).

The next week, more members approached Jacob with the same request. He asked where they met and when (he knew full well, having been briefed by his predecessor). They responded—Wednesday night, at the little school auditorium—and he suggested he might show up.

But he didn't, so back they came, more of them and more insistent. He had to come; he had to meet with them; they needed him; they wanted him. And so in his fourth week in the area, Jacob showed up at the school on time, as did I. Approximately ninety youths, ages 11 to the early twenties, were waiting for him. Seniors, juniors, unborns, and two girls' groups as well were ready for their meeting and their new sponsor. Jacob had his gang, twice as large as that reported by his predecessor. He had something special to offer them, a warm and friendly and accepting and supportive adult who seemed to care about them. They weren't used to that, and they responded. Gang regeneration at its best: the building of gang cohesion.

5. Jacob's style presented yet another illustration, this time toward the end of his assignment to this same group after two and a half years. Within a short span of time in the early spring of 1965, several older members—seniors—were released from the secure facilities of the California Youth Authority and returned to the street. Jacob talked with them as they reappeared—they could make or break his operation, he felt: shades of Ranger—and he was impressed with their seeming maturity and desire to stay out of trouble. He

thought perhaps he could use this newly emergent sincerity from a group of respected seniors and so decided to cut back his own activities significantly.

He held fewer club meetings and for shorter, only *pro forma* purposes. He reduced the emphasis on dance or car-wash preparations and pickup sports activities. Figuratively, although not literally, he reverted to sitting in the window. Our street observations for those last few months of Jacob's tenure showed a measurable decrease in gang cohesiveness. It wasn't the dramatic drop shown by Manuel's group, but it was clearly there in the data and visible to the observational team as well. On his own, Jacob initiated the decline seen earlier in those gangs whose workers were transferred and in Manuel's Operators. Four gangs out of four. In each case, the affiliated girls' groups also either lost members or disappeared. The pattern was inescapable.

6. Finally, there was one example that was independently based on data and did not rely on our research team's field observations. Our delinquency data at the end of the project shocked the workers (who then rejected our data, as might be expected), caused consternation among their agency administrators, and fueled the opposition of the police. Using officially recorded arrest data over the four years of the project, we found a significant increase in the offenses committed by gang members. The increase included both prevalence (more gang members, therefore more offenses overall) and individual offense frequency (more offenses per member, on the average). This is a devastating finding, of course, for any crime prevention program, but in no way unusual as students of criminology will recognize.

This project-related increase becomes more important to my message when it is broken into some of its components. Greater offending was associated with the increase in cohesiveness in the worker-transferred gangs before these transfers but was followed by a decrease following their transfer. It was associated with the greater cohesiveness of Manuel's and Jacob's gangs until they cut back on their group programming. Offending decreased in Manuel's case in direct proportion to the decrease in cohesiveness. In Jacob's case, the change in cohesiveness came too late to permit measuring any offending change by the end of the project.

Finally, in carrying out the statistical analysis of the recorded arrests, we distinguished between "low-companion" and "high-companion" offenses. Some crimes tend to be committed by one person alone, or perhaps with one or two companions only. Others are more commonly committed in the company of others.[31] We separated the two types, although they are by no means without some overlap, because it might provide another test of the cohesiveness notion. Our hypothesis was simply that gang intervention, whether positive or negative, would have more effect on the high-companion offenses than on the low-companion offenses.

And so it turned out. When comparing the gang member offenses before and during the project period, we found no significant difference for the low-companion offenses. They remained stable. However, there was a very significant increase during the project among the high-companion offenses. It was this latter effect that produced the overall finding of increased crime during the project. It is pure speculation, of course, to say that this is because of the demonstrated increases in cohesiveness, but it nonetheless is informed speculation that fits in the overall pattern of findings I've been describing. It makes conceptual sense and has what social scientists call *construct validity*.

One can add to this picture from the Group Guidance project three additional, supportive items. First, the Roxbury Project carried out by Walter Miller produced data strongly suggesting the same cohesiveness pattern. Second, the analysis of the Chicago Youth Development Project by Hans Mattick and his colleagues also indicated the same pattern. Neither project was initially designed—as was our analysis, by contrast—to contribute to this question. Instead, the support emerged on its own. Third, my Ladino Hills Project with the Latins, mentioned in Chapter 1, was deliberately designed to test the notion that a reduction in cohesiveness would lead to a reduction in delinquency. It did. The frequency of offending remained constant (rather than increasing, as in the earlier project), and the prevalence of offenses dropped dramatically in association with about a 48 percent decline in cohesiveness. Thus from many angles, the case has been made.[32] We'll need to recall all this during the discussion in Chapter 5 of gang control programs.

3

Historical Background

*Sometimes I think our job is to take these ghettoized, antisocial black
gangs and turn them into middle-class, prosocial white groups.*

Street gang worker in Los Angeles, following the suspension of his
program as a result of police pressure

Because many of the earlier data on gangs were accumulated in
direct conjunction with street gang intervention programs in the
1950s and 1960s, they were tested in a crucible far more demanding
than today's ethnographic approach to gang research. Understanding
gangs as they were then is critical to understanding them as they are
now. This chapter discusses earlier periods, and Chapter 4 describes
the present. I'll spend some time in Chapter 3 on

- Earlier sources of gang knowledge, mostly those involving
  intervention programs.
- Principal categories of gang knowledge obtained.
- The kinds of youths who joined street gangs.
- More coverage of gang cohesiveness.
- The positive functions of street gangs.

A Brief History

In the early 1960s, two white rookie Los Angeles police officers cruising the West Adams gang area spotted a large group of black youths
moving south from the vicinity of a movie theater toward a commercial area. It was dusk, and there was no school activity to account for
such a group, so the officers confronted the group, put them up
against a wall in standard search procedures, and called for backup

and transportation. The suspects were arrested and booked for "gang activity," a legal if amorphous charge later taken out of the penal code as unconstitutional.

As it turned out, the two officers had made a bad mistake. Their "gang" was part of a movie matinee audience leaving the theater at the end of the show. None were gang members, but they were the sons of middle-class parents who sought counsel from the American Civil Liberties Union (ACLU) and sued the city for false arrest. In the first court hearing on the case, an ACLU attorney verbally tore the poor LAPD rookies into shreds: How long had they been working a gang area? How much training on gangs had they been given in the Police Academy? What were the typical symbols of gang membership? When and where are large gang gatherings likely to take place? What clothing, tatoos, or other gang paraphernalia had they seen on these youngsters? Do gang members typically walk in large groups? Did the officers note that there was a movie theater at this location? Had they heard of matinees? On and on and on, until finally the judge, without being asked to do so, summarily dismissed the charges against the youngsters. The inquiry was painful to watch. The civil suit, by the way, was won by the parents, and the city paid dearly for the failure of its agents to know the difference between a gang and a group of moviegoers.

The advantage, in those days, was all on the side of the defense. There was little valid information about gangs; police gang intelligence was minimal and not available to rank and file officers. Knowledge of gangs was available, but not widely so. This may seem odd, since references to gangs or ganglike groups have been with us for centuries.

According to *The Tale of the Heike*, an epic description of warring groups in twelfth-century Japan, it was common, on the brink of battle, for these warriors to shout out to their adversaries the glories of their clan and the deficits of their rivals, much as modern gangs may punctuate their drive-by shootings with shouts of self-glorification and derisive epithets for their victim-rivals.[1] In any case, ganglike behavior was not invented in the modern American inner-city barrio.[2]

The term *gang* itself has been traced back at least to the writings of Chaucer in 1390 and appears in Shakespeare's *The Merry Wives of Windsor*: "There's a knot, a gang, a pack, a conspiracy against me."[3] There were named gangs in London in the seventeenth century, and in 1925 the infamous "Forty Thieves" gang reigned in lower Manhattan. More modern times have seen the *halbstarke* in Germany, the *blousons noirs* in France, the "teddy boys" in post–World War II England, and the *bezprisornye* in postrevolutionary Russia.

But whether it was in the time of Chaucer, the Forty Thieves, or the *bezprizornye*, little was or is known of these ganglike groups. They were the subject of unsystematic writing, not systematic research. The two LAPD rookies had inherited the gang mythology of *West Side Story*. Their fate at the hands of the ACLU would be quite different today because the police now have much better gang information available to them, and the arrests would not be made.

The Explosion of Research on Gangs

Of all the reliable knowledge we now have about gangs, I'd guess that more than half comes from projects carried out roughly between the mid-1950s and the late 1960s. But how we got to that knowledge explosion is an interesting story in its own right. There was a rapid development of theories directly applied to juvenile gangs (the common term in the 1950s) that preceded by just a few years the development of public policy concerns with gangs. As the two, theory and policy interest, melded, they in turn led to and largely determined the shape and scope of some major gang research and intervention projects. From these latter, in an uncoordinated yet complementary fashion, came the explosion of gang knowledge.

I cite this concatenation of developments because it is common to decry the failed marriage of theory, research, and public policy. The gang era of the 1950s and 1960s is a most notable exception, however. I have no need to review this material in detail here—it's available elsewhere, and my purpose is to provide background, not a text.[4] But a brief description is very much in order.

The pertinent theories came in two quite contradictory forms, one predominant and one less so, but no less intriguing. The first is best located in the writings of Albert Cohen and a bit later in the work of Richard Cloward and Lloyd Ohlin.[5] Cohen's *Delinquent Boys* was the product of the Chicago setting, whereas Cloward and Ohlin's *Delinquency and Opportunity* had a New York City context. Both are examples of what came to be known as *strain theory*, the notion that lower-class, urban delinquency was a reaction to the strains of accommodating (or failing to accommodate) middle-class norms and values.

Cohen's boys—gang boys in particular—were unhappy, frustrated youths, striking out against the "middle-class measuring rod," seeking in one another's company a common process for turning disadvantage into advantage. Cohen's boys were not particularly likable, being egocentric and hedonistic. Unfairly handled by American social structure, they were the antagonistic products of social strain.

Cloward and Ohlin took the argument further, locating the strains not only in the blocked opportunities to the accoutrements of middle-class goals but also in the strains of inadequate access even to illegitimate means to achieving those goals. The Cloward and Ohlin gangs thus became more rational than Cohen's because they sought out alternative means to personal and group satisfaction. These means were, most notably, theft, conflict, and the retreat into alcohol and drug use. Again, these are not happy youngsters. Social structures act to defeat their advancement, and the adoption of various gang cultures becomes their adaptation.

The contrasting theory was offered by Walter Miller, whose work concentrated on gangs in South Boston. Whereas Cohen, Cloward, and Ohlin had their ideas shaped by structural sociological thought, Miller, an anthropologist, was clearly affected by cultural relativism. For him, lower-class gang youths were not reacting against a middle-class source of strain; rather, they were enacting a set of "normal" lower-class values of toughness, fatalism, and "street smarts." They reflected their own situation rather than reacting against another. Indeed, Miller located some of the gang members' problems in the imposition of inappropriate conduct standards by the agents of middle-class society—the police, the teachers, the church, and agency leaders.

Miller's boys were the normal product of their setting, but again, they were not having much fun. Indeed, in the 1950s the plight of the Miller boys and the unfair situation of the social-strain boys of Cohen and Cloward and Ohlin called forth a social response of both empathy and sympathy. Their situation, however structurally derived, required attention not by changing them so much as by altering the structural settings in which they were forced to develop.

So the general theoretical tone had been set; the fault lay in the way that the social structure limited, blocked, and almost forced the development of opposing urban structures, the subcultures of gang life. At the same time, in New York, Los Angeles, and Boston, gang intervention projects had developed along lines suggested for Chicago by the urban theories of Clifford Shaw and Henry McKay and the observations of Fredric Thrasher. The notion was to enrich the local community's capacity to handle its own problems, including the recruitment of streetwise young men to work with the local gangs. These street workers became the heart of programs of the New York City Youth Board and the Group Guidance Section in the Los Angeles County Probation Department.

In varied forms, the street workers became the change agents and data sources for Miller's project in Boston, the YMCA program in Chicago, and the Boys' Club program also in Chicago (with Jim Short and Hans Mattick, respectively, as research directors), and my own evaluations in Los Angeles. All this came about because the logic of the underlying theories and the liberal tone of their intervention strategies appealed to two important and powerful audiences, the

Ford Foundation and President John F. Kennedy's Committee on Juvenile Delinquency and Youth Crime.

The conceptual leaps, along with the ongoing demonstrations of community intervention—especially via street workers—caught the imagination of Ford Foundation executives. They in turn wielded direct influence on Kennedy's committee, chaired by his brother, Attorney General Robert Kennedy. Woven in and out of these influences was the quiet but persuasive voice of Lloyd Ohlin, the Columbia professor with the winning idea. Ford and the "Feds" worked together to found projects in New York, Chicago, and Los Angeles, based principally on their acceptance of Ohlin's presentation of opportunity theory, as the Cloward and Ohlin work was known. The funding was, for those times, massive, supporting up to five years of intervention and data collection. Let me emphasize that intervention and data collection went together. The explosion in data on street gangs was made possible by the existence (sometimes forced) of collaborating street worker projects. The workers thus became the main avenue of access to the gang members.

With the New York and Boston projects already under way, the Ford–Fed alliance added Chicago, Los Angeles, and more projects in New York (Cloward and Ohlin's Mobilization for Youth Project on the Lower East Side). Much of what I will describe in the remainder of this chapter results from these various projects. They would not have evolved without the conceptual leaps of strain and lower-class theories; they would not have evolved without the liberal, social appeal of those theories in a liberal era of increasing federal concern with social problems; and they would not have evolved without the academic and social interests of the various project research directors—people like Walter Miller, Hans Mattick, Jim Short, and my own colleagues in Los Angeles.

Because I think it's important to give credit where it is due yet seldom accorded, I want to give special attention to the research component of the Chicago YMCA project and its direction by Jim Short. Sociology graduate students who study criminology quite uniformly read and can cite the theoretical works of Cohen, Edwin Sutherland, Thrasher, and Cloward and Ohlin. Many even read an article or two by Walter Miller. But when I raise the book *Group Process and Gang Delinquency* by Short and Strodtbeck, I get a sea of blank faces.[6] What a tragedy. Sociology students tend to read for ideas but not for data. *Group Process and Gang Delinquency* provides more valuable gang-relevant data per square inch than any other volume yet marketed. It is must reading for any student of street gangs. I can be very clear about the reasons for this:

1. Project gangs were not the only source of data: The book compares black gangs with white gangs and lower-class boys with

middle-class boys. This comparative emphasis does much to correct for the usual one-city setting of most gang research.

2. The data are derived from multiple methodologies. Systematic observations, interviews, and worker reports about the gangs and gang members combine to yield a comprehensive picture.

3. Before starting the project, a special conference of leading gang experts was held to help shape the theoretical issues that the project data might address. Thus informed, the project not only collected a valuable compendium of gang data, but it also tested the relative merits of the theories of Sutherland, Cohen, Cloward and Ohlin, and Miller. None fared very well, Cohen perhaps least well because his theory was the most testable. There's an irony for you.

If these theories failed to be persuasive when confronted by data, this should come as no surprise. They were "first efforts," usually derived from specific settings but not from systematic data collection efforts. They were not so much wrong as inadequate. I'm reminded of the week when my assistants and I suspended our operations and went into the field to observe gang members to decide which theory seemed most relevant. When we came back and compared notes, we concluded that we had independently seen "Cohen boys," "Cloward and Ohlin boys" and "Miller boys." Furthermore, some gang members were exemplars of one theorist at one point in time and of another soon afterward. The gang world out there is far too complex to be captured by any one, nascent approach. The final reason to praise the Short and Strodtbeck volume is its clear, data-based demonstration of this now obvious conclusion.

One final word before moving on to a description of the gang situation in the 1950s and 1960s: The various research projects I've mentioned have made it clear, in retrospect, that gang knowledge should not be just city specific—we can now get well beyond Fredric Thrasher in Chicago, Lewis Yablonsky in New York, Walter Miller in Boston, or Malcolm Klein in Los Angeles. There are both city-specific data and generic data about gangs. Our expertise in the field of gang knowledge must encompass both and increasingly does so. Where we continue to fall down is in educating the interested lay public in these matters. Here we err in leaving it to the media, and what a mistake that has turned out to be.

Most serious writers about gang affairs know better than to rely heavily on media reports.[7] Such reports are no substitute for research, or even for personal observation or experience. Headlines seldom derive from or reflect careful evaluations of the available information on gangs.

Reporters for the *Los Angeles Times* recently took a most singular and painful look at their own paper's treatment of crime topics.

Reporter David Freed is one of several *Times* writers who have provided noteworthy exceptions to my comments about the media's handling of gangs and crime. In one article, Freed points out that the paper carried stories on 411 shooting incidents in 1991, involving 615 victims, but that incidents involving an additional 8,000 victims did not find their way into print. In other words, a selection process results in the reporting of less than 8 percent of all shooting victimizations.

Imagine the factors that lead to such (necessarily) extreme selectivity: multiple versus single victim, age of victim, race of victim or neighborhood of the incident, brutality of the attack, lethality of the weapons used, and so on, including, of course, evidence of gang involvement. There is simply no way that media reports can fairly represent the quantity or nature of gang crime, to say nothing of gang life generally.

Several years ago an intergang shooting in Westwood Village, an upper-class shopping area bordering the UCLA campus in Los Angeles, led to the death of a young Japanese-American visitor to Westwood. This killing of a totally uninvolved and innocent woman in true suburbia was given enormous play in the press and on television. Gang shootings had briefly escaped from the hidden ghettos. The whole city seemed shocked by these media reports which went on for many days, as did the reaction of the local West-Side city councilman who offered $25,000 for information leading to the assailant.

But where was the hue and cry over the several hundred other gang killings in that year? What other victims elicited such a munificent reward? The whole affair became an affair because of the media's response. It was the media that suggested that the killing was noteworthy because of the victim's race, class, and location. In the history of Los Angeles, I do not believe any gang killing of a black or Hispanic has brought anything like the same attention, or the same nonsense, to the nature of gang violence.

Information on Gangs from the 1950s and 1960s

I noted earlier that most of the gang activity in the 1950s and 1960s was limited mainly to a few major metropolitan areas: New York, Philadelphia, Boston, Chicago, San Francisco, and Los Angeles come most readily to mind, mostly because of research associated with intervention programs in those cities. Seattle, El Paso, and San Antonio also had some active, if small, gang intervention programs under way. Los Angeles, in particular, must be understood as meaning the metropolitan area, since gang activity was to be found in a dozen neighboring cities, including Riverside, Ontario, Fullerton, Compton, San Gabriel, Oxnard, Garden Grove, Pasadena, San Fernando, El

Monte, Westminster, and Anaheim. A look at a road map of Southern California will reveal how widespread gangs were even at that time.

My phone interview survey, described in the last chapter, revealed that other communities had street gangs as well, but they weren't recognized because no research was carried out there. A total of 29 of the 261 gang cities covered in my survey reported gangs before 1961. Included in these are 12 in Southern California and 8 in Arizona, New Mexico, and Texas. Thus 20 of the 29, or 69 percent, were in the southwestern corner of the country. They were cities with primarily Mexican-American gangs, including the traditional gang structure described next.

What were these gangs like in the cities without research? That's hard to know. With the exception of El Paso, San Antonio, and Seattle, whose program descriptions suggest the existence of the familiar, traditional gang structure, we can't recapture what wasn't documented in the first place. Police experts responding to my survey in 1991 were most certainly not experts in 1961; some weren't even alive then! So what follows in this chapter is based mostly on gangs and gang programs with research components attached to them. Most of these gangs were of the traditional form, with age-graded sub-groups and multiple-year histories. The reader is therefore duly notified that these gangs have an unknowable relevance to unstudied gangs and to gangs in unstudied cities in those earlier days.

Gang Structures

Concern with gang structure yielded several attempts at description. Yablonsky described his violent gangs as very large, swayed to acts of extreme violence by sociopathic leadership. Although most research has since discounted much of his description, it failed to recognize that he was describing a temporary anomaly rather than a generic form of gang. Yablonsky's observations took place in connection with only a few structures with very short durations (a matter of months) at the peak of their violent outbursts in New York. Nonetheless, his imagery regarding the "violent gang" has remained a dominant perception for the general public and public officials alike.

Similarly, the "supergangs" of Chicago—the Blackstone Rangers, Vice Lords, and Disciples—have captured public attention, even though they were geographically anomalous. They lasted for many years as amazingly large and effective confederations of local gangs, even organized enough to siphon off millions of dollars in federal community action grants. They developed coordinated chapters, commercial enterprises, and political affiliations, along with written oaths and constitutions. J. Michael Olivero has provided a recitation of supergang incidents, year by year, and their involvement in violence and drug sales control.[8] These supergangs seem superficially to have modern counterparts in the People and Folks confederations in present-day Chicago, but supergangs have not appeared in other cities, although the now-defunct Slausons, a loose black gang confederation in Los Angeles, took on the superficial look of a supergang in the 1960s.[9] I have a taped interview with a Slauson leader in which he names and locates over a dozen affiliated Slauson subgroups.

More numerous—but how much so we can't know—were the "spontaneous" gangs that sprang up for a matter of weeks or months and then disappeared. They tended to be small in size, perhaps twenty or thirty members, youngsters presumably "trying out" gang life but finding it inimical to their well-being. Such groups tended to be quite territorial but loosely organized, primarily male, and usually but not always composed of minority youths. They were, more than traditional gangs were, limited roughly to the adolescent age range.

The traditional or "vertical" gang was the one we learned the most from. Typically it had a duration of many years, in some cases regenerating itself over decades. Because of its stability, the traditional gang became the ideal target for gang research. One gang worker, assigned to the infamous Hazard gang in East Los Angeles, described to me a young boy beginning to hang around with Hazard gang members, even though the worker saw little need for him to do so. The youngster was a good student and had good personal skills and no prior contact with the justice system. The worker went to the boy's home to speak with his parents about avoiding gang affiliation. The parents were not at home, but the boy's grandfather was. When the worker made his pitch about avoiding gang entanglement, the grandfather rolled up his sleeve to display his Hazard tattoo with pride: What's good enough for me is good enough for my grandson, was the message.

Bob Baker, of the *Los Angeles Times*, provides another illustrative anecdote:

> In Pomona the other day, Leo Cortez, an East Los Angeles gang member of the 1950s who is now a county Probation Department gang worker, was driving to a junior high school, hoping to steer a 12-year-old away from gangs. Cortez figured he had an empathic edge. "I used to sell heroin with the boy's great-grandfather," he explained casually.[10]

The traditional or vertical gang was (and still is, as I shall relate in Chapter 4) characterized by having several subgroups based on age. They might be called seniors or *veteranos* or Old Heads—more recently, OGs for Original Gangsters—at the oldest level. Juniors or some other designation constituted a middle-age group, and midgets or babies or some other name would be at the bottom. My favorite, the Senior Gladiator/Junior Gladiator/Del Viking structure, had a group of 10- and 11-year-olds called "the Unborn Vikings." This age-graded structure of cliques is the single most characteristic feature of the traditional gang and the feature that also led it to be termed a *vertical gang*.

Quite commonly, this cluster of three or four male subgroups is joined by one or two girls' groups, semiautonomous units often originated by sisters and girlfriends of the boys but then augmented from among their own female friends. I'll cover the girl gangs later in this chapter—at this point, just keep in mind that they are more common than might be thought.

In the earlier period of the 1950s and 1960s, the age ranges of traditional gangs generally stretched from around 11 years of age to the early twenties, with the peak at around 17. Thus there were few very young members and few older hangers-on. Leadership was closely related to age, but not in the fashion usually depicted by the police and the media.

Gang leadership was (and is) highly varied, so generalizations are a bit hard to make, although I'll attempt some in a bit. What is clear, for the traditional gang, is that there was leadership at each age level. That is, rather than the leadership residing in a few selected older members, in the style of *West Side Story* and other fictional accounts, it was distributed throughout the age range of the gang. The youngest clique had its own leadership, as did the juniors, the seniors, the girls' groups, and so on. Older clique leaders usually had a stronger reputation, but it was not always the case that this reputation could extend to leadership over the younger members. Group process is far more complex than that and far more varied than the usual police or media stereotype indicates.

Another feature of the traditional gang, although one that can be applied to just about any street gang, is the distinction between core and fringe members. Essential to appreciating the character of gang membership is the following:

1. Core members—up to a third or even half of all members, not just an inner "hard core"—are considerably more involved in crime than fringe members are, although the two patterns of crime are much the same.
2. Core members are much more active in a wide variety of group activities than fringe members are.
3. These two factors, crime involvement and activities involvement, can be quite independent, so that one might become a core gang member through activities yet not necessarily be heavily involved in gang crime.
4. Core and fringe levels of membership do not seem to be related to standard demographic or sociological factors—age, family status, economic level, parents' education, immigrant status, and so on.

Knowing a member's placement on these dimensions is not enough to identify him as either a core or a fringe member. Rather, the distinction seems to be based more on psychological and social factors. Core members have more character deficits (e.g., lower measured IQ, greater impulsiveness, fewer social skills) and more need for group affiliations. That is, they are more dependent on their peers.

Gang membership, then, can be seen as more functionally important to core members, as the gang is serving needs not met elsewhere. Vigil suggests, for instance, that the "group ego" substitutes for the weak individual ego-identity. An important if obvious implication is that the fringe member is more easily weaned away from the gang than the core member is. Purple Pedro was a core member of the Latins, very active in a variety of group and criminal activities. Over a year's period, he was given ten jobs, two drug and psychiatric treatment sessions, one tatoo removal appointment, and innumerable hours of individual counseling, yet eventually he killed an innocent older man as part of a fantasized gang "payback" shooting. He went to prison and killed a guard while there. His commitment to the group defeated the most dedicated attempts to wean him away.

With this overview of subgroups, ages, leadership, and the core–fringe distinctions, we can look now at Figure 3-1 for an illustration of how this all comes together. Figure 3-1 depicts the internal structure of the Latins over six months of observation. Similar depictions of other Los Angeles gangs would yield essentially the same pattern, as would those described by researchers in New York, Boston, Chicago, El Paso, Columbus, Milwaukee, and so on.[11] In Figure 3-1, the black dots are core members, the white dots are fringe members, and the lines connecting them are lines of association—who is seen more often with whom—as determined by research observations over the six-month period.

The ages along the side of the figure reveal how few very young and very old Latin members were active during this period, with the largest number of members—the white and black dots—being in the 17- to 18-age range. Note how members are connected, forming cliques in the various age ranges. Note also that most members of large cliques are core members—black dots—and that fringe members most commonly appear as loners or members of simple pairs. Note, finally, that in the most active subgroup, at ages 17 to 18, the numbers are large enough to separate them into three subcliques. The one on the right consists of all the school attendees, the one on the left a clique of members residing within a block or so of one another, and the one in the middle comprising the most criminally oriented, "hardcore" members. Purple Pedro, our ten-job murderer, mentioned earlier, was one of those in the middle clique.

The point about age-specific leadership is fairly clear in Figure 3-1; there is neither one nor a few clear leaders. Highly connected members are not well connected to cliques above or below them. There is no organizational chart here, and even though some gangs may have "presidents" and "vice-presidents" and "war lords," the titles do not easily translate into influence throughout the gang structure.

Finally, one last general point about Figure 3-1: Consider the concept of gang cohesiveness. For those who portray street gangs as highly cohesive groups—for example, those who maintain that the gang is a fine vehicle for drug distribution—there is little confirmation in Figure 3-1. Rather, what we see is a rather amorphous collection of subgroups, cliques, pairs, and loners. This is not the picture of a tight structure. It's loose and somewhat fragmented, despite each member's commitment to the group and the submersion of his own identity into that of the group.

This is not an organization that can readily act as a unit. Indeed, I've seen enough internal fighting to wonder on occasion how these young people do manage to stay affiliated. But—and here's my warning to those who would attempt to intervene in gang affairs—a structure like this can become far more cohesive if it is led to or is forced

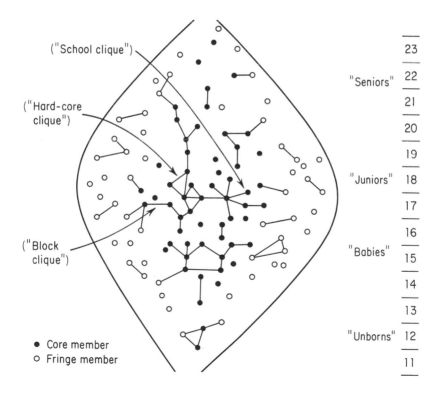

Figure 3-1. The Structure of a Traditional Gang. (Reprinted with modification from Malcolm W. Klein, *Street Gangs and Street Workers*, 1971, p. 66, by permission of Prentice-Hall, Inc., Englewood Cliffs, New Jersey.)

to. Rival gangs and police suppression programs can bring cliques together, force those loners into closer ties with their neighbors, and even link age-graded subgroups into more structural hierarchical relationships. The potential for increasing gang cohesiveness is in its normally moderate levels of cohesiveness. The consequence of an increase, we have learned, is not desirable.

Now, having provided these general points about traditional gangs of the 1950s and 1960s, let me move to some more specifics about gang leaders, gang girls, and gang territories. Then I'll follow with brief descriptions of other gang factors that became clear in the research of the 1950s and 1960s: crime patterns, ethnicity, member pathology, gang victims, and weapons.

Leadership

It is an almost impossible task, in a depiction of gang structure such as Figure 3-1, to determine who the "gang leaders" are, because leadership tends to be diffused over a number of members, even within each age level. Moreover, and not manifest in Figure 3-1, leadership varies among categories of activities; that is, a "fighting leader" may have little influence over social affairs or athletics or as a gang spokesman. In addition, the leadership shifts over time with changes in age, gang activity levels, and availability of members (owing to marriage, work, or incarceration, for example).

An interesting case is the member at the hub of the school clique. Luis (a pseudonym) was quiet and well liked by everyone. He was a favorite, furthermore, of the gang worker. Still in school, with a sense of personal responsibility for his behavior, Luis had been arrested only once in his eighteen years in the area, including four or five years in the Latins. The charge was for statutory rape, lodged by the parents of his girlfriend when she became pregnant.

The girlfriend gave birth to a healthy girl but died in childbirth. When her parents petitioned for custody of the baby, Luis contested the petition, seeking custody himself. He won. My point? Simply that here was a "gang leader" who certainly meets no stereotype available about gang leaders. Luis wasn't a highly visible fighter; he had almost no record in the justice system (though I've been assured that he was very active in delinquent episodes); he had a steady girlfriend; he was completing his high school education. In fact, his situation was strong enough to sway the welfare system in favor of his retaining custody of his child. Yet you didn't have to hang around Latin territory long to discover Luis's reputation as a Latins' "leader." Luis is more typical of a gang leader than are the many hard-core gang fighters usually depicted as gang leaders.

The stereotypical leader—the big fighter, the hardened criminal— doesn't retain his influence over many members, in my experience.

Nor does he maintain influence over long periods of time, not unless he's got a lot more going for him. Let me offer two contrasting cases, both from the same large black gang in Los Angeles. The occasion was a period of weeks in which the Red Raiders, the gang in question, was in intense conflict with the neighboring Victors.

One older member—call him Gregory—whose reputation derived strictly from his criminal orientation and his willingness to engage in violence, especially against other groups, monopolized a gang meeting discussion about retaliation against the Victors. The gang worker was urging mediation—a truce meeting to iron out the difficulties between the gangs. Slowly but surely the worker was gaining the day, aided by the elected president of the Red Raiders. Gregory, though full of bombast, was not swaying his audience of some fifty or so fellow members. Finally, in desperation, he stomped toward the door, signaling others to follow and proclaiming, "Ya win the last fight; then ya conversate!" He barged out the door; no one followed him.

When I talked with the Red Raiders, they generally agreed that Gregory was a "leader": He had influence; he was a senior who had been a core member over many years; he had a "rep." But as the incident showed, leadership in the gang can be very situational. The police knew all about Gregory and had put him away on more than one occasion. They knew him as a Red Raider leader. Yet putting him away had virtually no effect on the Raiders. Leadership is a group function; it doesn't simply reside in an individual person.

My other Red Raider example comes from the same intense period, but another even larger meeting. The president (who had succeeded his elder brother in that position) was cautioning his group about their concern with fighting the Victors when they should be thinking of their futures. The president had been highly influenced in these matters by his gang worker, among others. The Red Raiders, he maintained, should be thinking about giving up this gang life. After all, he was working, he was going to night school, he was straightening out his life, and they should be doing the same.

Indeed, as I listened to this rather remarkable talk, which went on for close to twenty minutes, it struck me that the president was doing a better counseling job than I'd ever heard from the gang worker over a three-year period, or from any gang worker, for that matter.

But finally, the president's capital was used up. The group had gathered to plan their response to the Victors, not to plan their lot in life. The undercurrents of impatience and excitement slowly but very visibly, I thought, were coming to the surface, until from the back of the room came the direct challenge to the president, shouted out to an echo of appreciative "yeahs" and "right ons": "What we gonna do, man, if them Victors come over onto our set one more time?"

The president's response to the challenge came without hesitation, without equivocation, and notably without any relationship to his impassioned plea for self-betterment. "If they come over here," he shouted out, "we gonna go through them like a hot knife through butter!"

The response was something to watch. A sea of bodies, maybe seventy of them, shot into the air, arms waving and hands clapping, shouts of slogans and obscenities attached to the word *Victors*, feet pounding the floor—it wasn't at all clear to me that the building could handle the acclaim. With one deft phrase, properly delivered and timed, the president had replaced his counselor role with his fighting-gang leadership role. His reputation was not only saved, it was clearly enhanced. My point, again, is that gang leadership is functional, not just positional. It exists more, and less, because of its context. My examples, in line with this chapter, come from the 1960s. My conclusions, as far as I can learn, hold up almost as well today as they did in the 1960s.

Walter Miller noted the tendency for casual observers of gangs to attribute one of three leadership stereotypes to them. The first is a "military command" model; the second is a "key-personality" model; and the third is the "collective leadership" model. After describing them, he comments correctly, I believe:

> The three leadership models described here are pure types; many gangs incorporate aspects of two or all three. Conventional conceptions tend to favor the military-command and key-personality models, with many observers, including practitioners, viewing gangs as autocratic despotisms or highly disciplined hierarchies. In reality, leadership in the average gang is far more flexible and democratic than the military and key-leader images imply, although some gangs closely approximating these two models do exist. One possible reason that gangs with central-ized authority are seen as more prevalent is that they appear to be more amenable to methods of gang control frequently advocated by the police and the courts. If one attributes crime by gang members primarily to the presence and influence of a single strong leader (key-personality model), all one has to do is to locate and remove that leader to curb or eliminate crime. Similarly, if gang leadership takes the form of explicit offices, such as Prez, Vice Prez, Warlord, and so on, those who hold such positions can be readily identified, and removed from their positions by arrest and incarceration.
>
> Since gangs whose leadership is as concentrated and coercive as is implied by the centralized-authority models are probably a small minority of all gangs, the fact that they are seen as the most prevalent type is prob-ably due at least in part to programmatic concerns rather than considera-tions of factual accuracy.[12]

Girls

In most street gangs in the 1950s and 1960s, girls did not normally participate in the formal gang meetings that took place during intergang hostilities, although they certainly were present at many other meetings and activities. These exclusions were quite deliberate, and for two reasons. One is pure chivalry, and the other is the concern that girls would "fink," tell what was going on to those who shouldn't hear. These were the findings in a series of interviews with male gang members about other planned criminal events. They didn't like to see their girls get in trouble, and they didn't trust the girls to keep quiet about the plans. In addition, some of the boys felt they would lose respect in the eyes of favored girls if they were apprehended for some crimes.[13] Property crimes were the least likely to involve female companions, perhaps because property crimes more often involved planning, that is, the time for exclusions to be exercised. Anne Campbell reported the same male gang members' perspective in New York: "Though women may be fairly loyal to their male partners, they cannot be trusted to hold group secrets, and so they are excluded from the planning and performance of crimes."[14]

It was rare in earlier decades to find truly autonomous female gangs.[15] It was also rare to find gangs in which girls participated equally with boys. One fully integrated group existed in Honolulu during the late 1960s in which 36 out of 66 known members were female.[16] Rather than being either autonomous or fully integrated, most female gangs had an "auxiliary" status. They were semi-independent, often with their own names, yet dependent for their continuation as a group on the connection to their male counterparts, as described earlier. Of forty-seven cliques in two Mexican-American barrios studied over a forty-year period in East Los Angeles, fifteen were female groups.[17] The ratio approximates what existed in black gang areas in the 1960s. Such groups had names of their own or feminized versions of the male gang names—Gladiators, Gladiettes; Locos, Lil Locas; Businessmen, El Social Playgirls; Bartenders, Bartenettes. My favorite name was the girls' group associated with the Del Vikings. They called themselves the Del Vi-Queens, a linguistic perversion that never failed to amaze my Scandinavian criminology hosts.

The girl gang could be found in most gang-involved cities (including Boston, New York, Chicago, Los Angeles, El Paso, and Philadelphia) and ordinarily in the auxiliary status I've described. Although these reports suggested a male-to-female prevalence rate of anywhere from three to one up to ten to one, the 1980s and 1990s reports suggest that this may have changed. Reports from Rochester, Denver, and Pittsburgh, for instance, indicate more even ratios. Since these are based on self-admission, compared with outside observations in the earlier years, it's hard to distinguish a change in prevalence from a change in the method of collecting data.

Another thing to note about the girl gangs, just as with the boys, is that they came in a wide variety of sizes, structures, age ranges, and levels of criminal involvement. The best generalization is not that they had a particular character but that they showed considerable variation on just about all dimensions of interest. Some had only a half-dozen or so members, but they could go up to thirty or more. Some had rather clearly defined membership, and others were essentially leaderless and therefore more dependent on the boys. Some were capable of extreme violence, and others none at all. But all shared the fate of the gang world—the willingness to become criminally involved, the susceptibility to victimization, and the potential damage to their own adult careers. Girl or boy, it didn't matter: Ganging was often a self-destructive pattern over the long run.

There also were more specific similarities between female and male gangs. Although younger than their male counterparts, the girls presented similar demographic profiles in terms of family construction, socioeconomic status, and reasons for joining gangs. They also seemed to share similar delinquency patterns, although at considerably lower levels. They were also as sensitive to their own gender peers as were the boys to theirs. It was not the case, as suggested in some loose journalistic writings about gang girls, that gang girls were normally the sexual playthings of the male gangs or the secret weapons-carriers for warring male gangs. Instances of both patterns tended to be blown way out of proportion. If gang girls were promiscuous—and many of them most clearly were—that promiscuity was as likely to precede as to result from gang affiliations. If they were prone to delinquency, that too was a factor for gang selection, not something merely forced on them by overbearing male gang members. Girl gangs did not exist solely to serve the needs of male gangs. The girls were fully capable of creating their own versions of ganging.

I'm reminded of one instance, fully aware that one case cannot yield a generalization, of the formation of a new and autonomous girls' gang. It happened because an energetic and assertive leader of an established group was ousted from her position by a small group rebellion. Bonnie had been using her group to feed her own very strong needs for recognition and status while losing touch with the needs of her friends.

Her response was to wander a dozen blocks west to a playground that had no gang tradition but nonetheless was a gathering spot for local kids on the loose. There, she formed out of whole cloth a totally new, albeit short-lived, female gang, the Bonnettes, independent of male gangs and reflecting her views of what a girl gang should be. Her own social pathology is one part of the story; the susceptibility of unattached local girls is a second; a third is the total independence of the gang's development from male gang structures.

Unlike the Bonnettes, girl gangs arise for the same kinds of social structural reasons that male gangs do. The institutional disorganization of the inner city, the family, school, health, and employment context is where the real distortions lie. These groups are a response to these social distortions, and seeking individual pathologies to explain gang joining is not a very fruitful pursuit. Indeed, the more pathological gang member is usually on the fringe, not at the core. Bonnie was ousted from her original group.

The only modification I'd offer is that females, being under such social pressures to develop into nice young ladies, must tread a fine line between their gang roles and the more traditional role behaviors for girls. The girls must find a form of "acceptably deviant behavior" in order to maintain their dual roles of gang girls and developing young women.[18] Some succeed far better than others do.

Finally, it bears repeating about female gangs that they are not simon pure. They did and do get involved in criminal activity, although many members may never develop an official police or court record. They do run away, they do steal, they do engage in illicit sex, they do fight. I've seen a videotape of two girls in a prearranged battle over the "rights" to a particular male gang member. The tape first shows a lot of pushing and shoving and hair pulling, then an unsolicited knee to the groin from a female spectator, and it finally ends after one of the two combatants pulls a concealed knife from behind her back and repeatedly stabs her opponent to the point of near death. It's not pleasant to watch; it's not a common event, yet it does illustrate the potential in any gang confrontation, male or female.

Regardless of the prevalence of arrest, most of the female gang crime consisted of minor offenses (drinking, sex, thefts, joy riding, and the like), and nonviolent offenses. It was varied in the categories of offenses, but not as threatening to future life or to the girls' careers as the male gangs' crime was. It deserved attention as part of the ganging process, but not because it had some sort of peculiarly female twist to it.

Crime

Let's get back, then, to the crime and delinquency associated with the male gangs in the 1950s and 1960s. I can offer some generalizations that stood up quite well across those cities that were observed. The reader must remember, though, that these generalizations mask a great deal of variation from gang to gang.[19]

Despite such variations among gangs and their criminal activity levels, the following seem to be true:

- Gang members were far more criminally active than nongang boys were, and more so than delinquent nongang boys as well. Some gangs, by the same token, were more criminal and some more violent than others.

- By and large, gang members showed very versatile patterns of criminal involvement, rather than specializing (as some scholars had thought) in theft, or conflict and violence, or drug use. There were some specialty cliques within some traditional gangs, but these were the exceptions, as multioffending members engaged in a wide variety of offenses—what I refer to as cafeteria-style offending.
- The bulk of the offending was in relatively nonserious categories: thefts, vandalism, status offenses, minor assaults, and so on. In both their cafeteria pattern and the preponderance of less serious offenses, gang members manifested a pattern common to most young lawbreakers—higher numbers but similar patterns. The media's portrayals of cohesive, violent marauders of innocent bystanders simply did not hold up to empirical analysis.
- Drug involvement (even excluding the ubiquitous use of alcohol) was fairly common, but seldom to the point of addiction. This involvement, it should be carefully noted, refers principally to drug use, not drug sales. Marijuana and pills were the principal drugs of choice; heroin was known but heavily used by very few gang members. Cocaine was all but unknown. A number of writers have noted that heavy users were often ridiculed or ostracized by other gang members. Addicts could not be trusted with gang plans or gang protection. Drug sellers were mainly doing friends a favor; they were not seriously in the business of selling drugs.
- Gang violence—especially serious violence in contrast with common street fights—was the more unusual behavior. Of many planned intergang "wars," few actually took place. The "rumble" usually fizzled. As Miller notes and I can attest, "One or both sides found a way to avoid open battle. A major objective of gang members was to put themselves in the position of fighting without actually having to fight."[20] To illustrate such a process, I offer the following incident I observed:

The Red Raiders were informed, after a series of minor skirmishes, that the Generals would be waiting for them at 3:15 p.m. outside a local junior high school. This was to be the decisive engagement. During a Raiders' meeting in which various countermoves were considered, the gang worker developed a perfect ploy. To meet the Generals at their appointed time and place, he suggested, was poor strategy because it was picked to favor the Generals. Instead, the worker suggested, the Raiders should choose their own time and place. He suggested 3:30 p.m. at a park in their territory about six blocks from the school. The Raiders accepted this idea.

Indeed, there they were, scores of them, ready to battle at 3:30 at the park. But of course, no one had informed the Generals of this alternative. So while the Raiders whooped it up at the park, noting that the Generals were "too chicken" to take them on, the Generals were doing the same at the school. Both sides, in their views, had saved face, and neither had to face actual combat. At about 4:00, before either side could plan a delayed march on the other, the police showed up, having been alerted and given the 4:00 time by the worker. No war, no serious loss of face, a chance to crow over the cowardly stance of the enemy, and an emotionally drained but happy gang worker were the result.

Aside from the issue of gang fights, however, there was nevertheless a worrisome level of potential violence in the gang world of the 1950s and 1960s. People did get hurt; people did get killed. While we were observing the Latins over two years, there were three killings. In the following three years, a dozen more went down, killed or overdosed. Drive-by shootings were not uncommon, even though today's police and writers believe that drive-bys are a new feature of gang life.

I have three taped interviews with gang members in the mid-1960s. All three describe drive-by shootings in a matter-of-fact way, a part of gang life. A killing that deeply affected the Latins' gang worker was a drive-by shooting in his presence; later, he himself looked down a pistol barrel aimed at him from three feet away as a gang car pulled up beside his. "Not a groove feeling," said one gang member about such a direct confrontation.

In 1968, I was informed that a Latin member was gunning for me, having been told that I had reported him and his brother to the police as suspects in a shooting. Who told him this, or why, I never learned. But I knew I couldn't continue on the project while swiveling my head around at every sound for fear that Leon was going to take me out. He, like many of his homeboys, had the potential for true violence, even if it were seldom realized.

Fortunately, at that time Leon was in grave condition in the hospital, having been deeply stabbed in the stomach at a party. It seemed there was no better (safer) time for me to deal with him. I went to his room and found him quite immobilized. We chatted about his wounds but, more important, about the time I got him enrolled at no charge in an art class in the vaunted Chouinard Art Institute and the time I had transported him and his goods some thirty miles to move in with his girlfriend. By the time we were through "bonding," there was no hint of reprisal for my rumored act of informing. Nor did I hear anything of this following his recovery. But let's be clear; as much as I have put down the reputed violence levels of gangs in those days, it was not without recognition for its potential.

Many people overestimated the level of gang violence in the 1950s and 1960s—and do so today as well—because they rely on media accounts in which reporters highlight the relatively uncommon violent acts and overlook the more typical, boring aspects of a gang member's life. A second source of overstating gang violence is listening to gang members themselves. Gang members *talk* violence a great deal; they *do* violence far less. A 1950s New York City Youth Board worker described the relatively mundane nature of gang life:

> As soon as the boys realized that I was not going to take out a badge and say, "Let's go" the first time one of them pulled out a pair of dice, I learned that such things as the hijacking of bakery and soda trucks, the rolling of drunks, and joy riding in stolen cars were their behavior pattern. Three boys were sent to jail for stealing and wrecking a car before I'd had a chance to know them very well. I was shown zip guns and was assured that they would work. I was invited to partake of some of the loot of a rifled candy truck. One night, one of the boys was waylaid and badly beaten by members of a rival gang.
>
> While the anti-social activities of the Boppers were dramatic, it should be noted that they represented a very small segment of the time and activities of the Boppers. Actually, the most significant characteristic of their lives was the fact that it was aimless, disorganized and unproductive.
>
> The average boy had no job and was out of school. He slept late, met his friends on the corner and moved back and forth from the corner to the candy store, the poolroom and back to the corner. He ate snacks with his friends in the luncheonette and returned home only to sleep, usually in the early hours of the morning.
>
> The fellows themselves were aware that their activities were limited and monotonous. One boy once told me, "Now, for example, you take an average day. What happens? We come down to the restaurant and we sit in the restaurant, and sit and sit. All right, say, er... after a couple of hours in the restaurant, maybe we'll go to a poolroom, shoot a little pool, that's if somebody's got the money. O.K., a little pool, come back. By this time the restaurant is closed. We go in the candy store, sit around the candy store for a while, and that's it, that's all we do, man."[21]

Ethnicity

In earlier decades, most street gang were composed of various white ethnic members. There were Irish gangs, Polish gangs, Italian gangs, and so on. By the 1950s and 1960s, this was far less true. In places like Boston and Chicago, some white gangs existed, but blacks and Hispanics (Mexican or Puerto Rican) had become the predominant participants. The southwestern gang cities were dealing with Mexican Americans, Philadelphia and San Francisco with blacks. In Los Angeles, Mexican-American gangs clearly predominated, but blacks were rapidly making their gang appearance. There was one well-known Asian gang and no whites. Indeed, the LA gang workers, almost all minority members themselves, were always on the lookout for signs of a white gang to help redress an imbalance they saw as denigrating to their own ethnic groups.

The Chicago data, which compared white with black gangs on a number of points, nonetheless revealed that these differences paled among the similarities attributable to class and gang context.[22] When expanding my own research operations from South Central Los Angeles (black gangs) to East Los Angeles (Hispanic), I also found the similarities to be striking.

The other reason for stressing cross-gang similarities is a bit more subtle. Ethnic gang distinctions can't stand out if generic gang variation is high across gangs. As Miller comments, there is a tendency "to conceive local conditions as unique. The gang situation does indeed differ in significant respects from one area to another, but is also alike in many important ways. A balanced treatment requires the identification of both generic, cross-community characteristics and local, individual characteristics."[23] We need to be able to describe a more generalizable gang picture along with the variations inherent in that picture. Miller, for example, concluded that "...core characteristics of the gang vary continuously from place to place and from time to time.... Gangs may be larger or smaller...modestly or extensively differentiated, more or less active in gang fighting, stronger, or weaker in leadership, black, white, yellow or brown, without affecting their identity as gangs."[24]

Pathology

A little later, I'll have more to say about why youths join gangs and what kinds of youths are attracted to membership. However, it's important to make a separate statement about the often-assumed, pathological nature of the gang member. Sparked principally by Yablonsky's book *The Violent Gang* (1963), considerable controversy emerged over just how "normal" gang members were. The general public, the media, and law enforcement officials were easily convinced that gang leaders were sociopaths who could influence overly susceptible gang members. Most scholars of this era, however, while noting that some gang members were seriously disturbed, concluded for the most part that gang membership was more a matter of social niches than of serious individual deficiencies. Individual pathology was given short shrift, and demonstrations of individual deficiencies among core gang members were largely ignored. Had violence levels been higher, perhaps this view might not have persisted.

I'm not saying, of course, that pathology was absent. Each gang I observed had one or more members that I judged to be seriously disturbed or often out of touch with reality—even his own gang reality.

One of them was the victim of excessive glue sniffing. Gregory, the rejected leader of the Raiders described earlier, was another. So was "Crazy Carlos" of the Latins, a heroin addict normally shunned by other members as being too unreliable yet capable of stirring the group to impulsive action in a crisis situation.

In fact, there were several members of the Latins for whom we wanted to arrange psychiatric consultations—Crazy Carlos, Purple Pedro (who went through so many resources and eventually murdered), and a third lad who had obvious periods of severe depression. We failed. A psychologist volunteered his time, only to find that he couldn't handle the open street setting that gang members chose, whereas a community psychiatrist thought the project's field staff was the better target for his skills!

For most gang members, though, therapy is far less important than education skills, job skills, and a chance to break out of the reliance on their peer group for ego satisfaction. Counseling may be helpful, but not in a context without alternatives. Most gang members are surprisingly close to normalcy, given their pathogenic settings in the ghettoized areas of our cities.

Victims

This description would be unbalanced if I didn't address the question of who else gets hurt because of gang predations. I say who else, of course, because gang members themselves and their families are clearly damaged by some of their criminal involvements. But if we consider the direct victims of robberies, assaults, homicides, and the like, when the characteristics of the victims can be known to some extent, what do we find—who are they?

Not surprisingly, the answer is similar to the one criminologists love to spring on their new undergraduate students in the first classes of the semester. The naive student supposes the victims of most person-to-person crimes to be the innocent, the defenseless, the elderly. But their criminology lecturer tells them no; the victims look much like the suspects. Both tend to live in the same neighborhood and to be of the same racial or ethnic status, the same social status, the same age bracket, the same gender, and so on.

So it was in earlier decades with gangs. Black gangs generally attacked black gangs, Hispanic attacked Hispanic gangs, white gangs attacked white gangs (though often of different national derivations; I have a Jewish colleague raised in the Boston area who claims that the Jewish gangs always won against the Irish gangs, and he can't be shaken from this memory). The most common gang victims are young males of the same race as the offenders. Both are probably inner-city residents, sharing many more characteristics than either would with their white suburban nongang metropolitan neighbors.

A study in Philadelphia by the Pennsylvania Crime Commission in 1969 found the victims of gang homicides to be gang members themselves in over 70 percent of the cases from 1963 to 1969. Eighty-eight percent were under 25 years of age, as were all of the assailants. In 1968 and 1969 when the data were available, 89 percent of the victims were black.[25] A study by Bernard Cohen in the same city revealed that gang victim–suspect similarities were even greater than victim–suspect similarities of other group but nongang events.[26] The gang exaggerates the in-group nature of the relationship.

A 1966 study of seven Boston gangs, using 125 violent incidents in which most of the gang assailants were adolescents, found that the gang victims were very similar to the assailants. Adolescents comprised 78 percent of all the victims, almost 60 percent being gang members. Only about 7 percent were of a different gender, and about 28 percent were of a different race. In other words, Boston gang members picked on their own kind—in several senses—rather than moving out to assault a contrasting enemy.[27]

The situation in Los Angeles was similar. To know the character of the gang members was to know the character of the victims in the majority of cases. It was often supposed that this was one reason for the minimal response from political leaders; for the most part, their constituents were not being victimized. The same attitude was attributed by some to the police, and in truth I did hear more than one cop, in more than one city, express the view that they'd just as soon see the gang members knock off one another. In the old days, that's just what they did, but not frequently. Nowadays, they do it more frequently, with the same target concentration. In 1981, the Los Angeles Sheriff's Department's data revealed 77 percent of gang homicide victims to be affiliated with gangs themselves.

Although firearms are frequently used in gang offenses today, this reliance on guns was far less true in the earlier decades. Every on-site description by gang researchers noted the wide variety of weapons used when the occasion arose—firearms, zip guns, knives and other sharp instruments, chains, clubs, tire irons, bottles, and a host of other weapons. And of course, many fights involved little more than the human body.

The Pennsylvania Crime Commission reported that 61 percent of the gang killings in Philadelphia in 1968 and 1969 were shootings. The comparable figure for a six-month period in East Los Angeles, as reported by the Sheriff's Department, was exactly the same, 61 percent. The contrast with the present is striking. Firearms are now standard. They are easily purchased or borrowed and are more readily available than in the past. Officially recorded weaponry in Los Angeles homicides has climbed steadily to 90 percent firearms in the last few years. Our own intensive file search put the figure at about 80 percent in the early 1980s. Although it is true that the weapons

Table 3-1. Number of Incidents with Presence of Other Weapons, All Samples Combined

Category	Incidents	Examples
Tools	95	Hammers, pipes, wrenches, screwdrivers
Body parts and clothing	94	Fists, feet, combs, neckties, shoes, belts
Alcohol related	89	Bottles (usually beer), mugs, glasses, bar stools
Auto related	84	Tire irons, bumper jacks, car as battering ram, hit-and-run cases
Neighborhood items	58	Bricks, boards, branches, asphalt chunks
Home items	55	Vases, vacuum hoses, forks, brooms, patio chairs, flowerpots
Sports related	51	Baseball bats, pool cues, barbell weights, golf clubs, ping pong paddles
Traditional and other	46	Blackjack, fire bomb, mace, milk crate, cane, handcuffs, laundry cart, cash register

have become proportionately more lethal as well (automatic pistols and rifles), normal handguns, rifles, and shotguns still comprised 89 percent of the firearms over the 1980–90 period in Los Angeles (though dropping to 80 percent in the last part of the decade). A similar overall estimate has been made for Chicago over the past fifteen years.[28]

It's of some interest in this connection to consider what alternative weaponry is at hand. We looked at this in our gang and nongang violent incidents in Los Angeles in the early 1980s and found the nonfirearm weaponry noted in the police investigation files (see Table 3-1).

I suppose the conclusion would be that where there's a will, there's a way. Almost every setting contains potential weaponry, and if absolutely nothing else is available, there's always the fist and the foot. I'd rather take my chances as a victim of fist or foot than of a shotgun blast.[29]

Who Joins Gangs?

Writing about Chicano street gangs, Diego Vigil captures nicely some aspects of how youths are drawn toward them. He notes:

Street socialization is an aspect of the barrio that undergirds established gangs.... The most multiply marginal individuals are often the most unsupervised and reside in crowded housing conditions where private space is limited. These youngsters are driven into the public space of the streets where peers and teenaged males, with whom they must contend, dominate. These peers and older males provide such youths opportunities for a new social network and models for new normative behavior, values, and attitudes. They also generate a need to assuage basic fears stemming from not wanting to be fair game for anyone.

Thus one of the first goals in the streets is to determine where one fits in the hierarchy of dominance and aggression that the street requires for survival. Protection comes from seeking associates who are streetwise and experienced and willing to be friends. In turn, this prompts the youth to return the favor by thinking and acting in ways that his friends approve. The new social bonds are reinforced, a sense of protection is gained, and new behavior patterns and values are learned.[30]

In this section, I want to bring together various writers' notions about who these gang members are. To set the stage for this discussion, I want to repeat that I'm concentrating here on what I prefer to call the street gang. I am not including motorcycle gangs, prison gangs, stoners, and organized crime groups. I am not talking about supremacist groups, skinheads, political terrorists, or militant activists. By the same token, I am not describing taggers, break-dancers, play groups—"that good old gang of mine," as the song goes—and all the many adolescent groups to which many of us belonged that did on occasion get into trouble with the law.

What's left? As noted earlier, groups of young people, who may range in age from 10 to 30 or occasionally older, whose cohesion is fostered in large part by their acceptance of or even commitment to delinquent or criminal involvement. They are principally but not exclusively male, principally but not exclusively minority in ethnicity or race, normally but not necessarily territorial, and highly versatile in their criminal offenses. These offenses are not predominantly violent, but they are disproportionately violent when compared with the activities of other youth groups or individual persons.

Equally important, in my view, is that street gangs are delinquent groups that have passed a "tipping point" in their confrontational stance as a group. They have set themselves apart from their neighborhoods in their own perceptions, and many members of the local community have come to see them as a group apart. In other words, the street gang comes to see itself as such, and so does the community.

In regard to who joins street gangs, then, first, it is not sufficient to say that gang members come from lower-income areas, from minority populations, or from homes more often characterized by absent parents or reconstituted families. It is not sufficient because most youths from such areas, such groups, and such families do not join gangs. There is a selection process that results in 1 percent, or 5 percent, or 10 or even 20 percent of gang-age youths choosing the gang option while the majority select themselves out.

Many years ago when I mounted the gang intervention effort known as the Ladino Hills Project, I was told by a juvenile division sergeant in the Los Angeles Police Department, a gang veteran who knew the area well, that "all the kids in Ladino Hills belong to the Latins Gang—all of them!" Yet over a year and a half, we ended up with a gang roster of just over a hundred Latin members, male and female, in an area that housed several thousand gang-aged youths. Our figure was similar to that of 123 members designated five years earlier in an intensive study by a local gang prevention program. So, in most instances, the gang itself is a minority within its own local age peers, many of whom share the same income levels, the same minority status, and the same family situation.

What characterizes the gang joiners and the heavier participators includes one or more of the following:

- A notable set of personal deficiencies—perhaps difficulty in school, low self-esteem, lower impulse control, inadequate social skills, a deficit in useful adult contacts.
- A notable tendency toward defiance, aggressiveness, fighting, and pride in physical prowess.
- A greater-than-normal desire for status, identity, and companionship that can be at least partly satisfied by joining a special group like a gang.
- A boring, uninvolved lifestyle, in which the occasional excitement of gang exploits or rumored exploits provides a welcome respite.

Gang members are not so much different from other young people as they are caricatures of young people. Their needs and their pleasures are exaggerations of those familiar to us from a more general youth population. Arnold Goldstein referred to them as *hyperadolescents*. It is not surprising, then, that one does not need stereotypical teeming tenements to produce street gangs; they can arise in the midst of a wide variety of inner-city, segregated communities.

Furthermore, it is not necessary—indeed, it is truly misleading—to apply to gang members such epithets as were used in a *Saturday Evening Post* article: "warped little creatures," "brats," "stunted personalities," and "misfits."[31] I knew some kids like that in my own school, and we had nothing even approaching street gangs.

Indeed, the problem with characterizing gang members in any definitive way is that they comprise such a wide variety of young people that every generalization will have its exceptions. I pointed out more than twenty years ago in *Street Gangs and Street Workers*: "Gang members come in all shapes. They are short and tall, bright and dull, aggressive and passive, easy to know and practically unreachable" (p. 81). Carl Werthman used as an example in his master's thesis the school performance of thirty core gang members:

- Four were fully illiterate.
- Twelve consistently achieved Ds and Fs.
- Fourteen consistently earned Cs or better, with four of these on the honor roll.

Werthman also described a lot of his observed gang members as highly aggressive, as a reaction to authoritarian fathers:[32]

> "He's a big one," one boy says. "He's got muscles on top of muscles. He threaten to beat the shit out of me—and he done it a couple of times." And another member testified to Werthman, "He used to whip me, throw me downstairs until I got big enough to beat him."[33]

The notion of this need to aggress is similar to the more recent portrayal by Sanchez-Jankowski that gang members—nearly all of them, as I understand him—are characterized by a "defiant individualism," so that the gang itself is a defiantly individualistic organization. The trouble with these depictions is that they simply don't apply across the board. My hunch is that these writers tend to think of their gangs in terms of the core members, those who certainly are more visible and more dramatic as individuals. But if you recall Figure 3-1, it depicted the gang with a lot of fringe members as well, those less likely to be aggressive or defiant or to participate in violent gang activities. Our descriptions therefore must encompass far more than this stereotype of postural (and actual) toughness, which clearly is there.

In a broad assessment, a Philadelphia gang research team applied a wide assortment of measured variables to gang members and non-members in that city.[34] They found a total of seventy-seven variables to distinguish gang from nongang, an enormous list suggesting that (1) the differences are indeed substantial and (2) there's plenty of room to encompass the variety of members I've indicated are out there. But few of those seventy-seven variables, on analysis, turned out to be predominant. In fact, only three require attention: self-reported (admitted) violence, parental defiance, and group rewards, such as status, companionship, excitement, and protection. These factors mirror, in a very gross fashion, the two factors reported earlier as distinguishing between core and fringe membership, greater personal deficits of core members and their consequent aggressiveness, and more group activity.

I recall vividly a setting in which a fringe member, a lonely and unhappy kid, was seeking more direct involvement in his gang. A core member, violent (there were many proofs of this) and committed to the group, openly resisted the fringe member's advances to the group. He initiated a truly vicious fight. An adult attempted to intervene to prevent serious damage but became caught between the two gang members who, trying for a death grip on each other, instead had a mutual death grip on the hapless adult. It was after about five, white-knuckled minutes of stalemate that the younger boy finally lost his grip and the adult was able to extricate himself from the vise, bruised, almost with the breath squeezed out of him, and knowing forever after that who joins and who does not join would be decided by the gang, not its observer.

Belonging—having the status of gang membership, the identity with a particular gang, the sense (correct or otherwise) that in the gang there is protection from attack—becomes very important, very rewarding to the member. It provides what he has not obtained from his family, in school, or elsewhere in his community. My colleague Diego Vigil wrote of the Chicano gang member that he replaces low self-esteem with group esteem. "Why we join, see," I was told by an early 1960s black gang member, "is because we wanna belong to something. We can't belong to the school band, to the school council."

For some members, this belonging was the supreme reward. A gang member may (though many do not) refer to his gang as his family. Journalist Léon Bing commented, "You know, gangs are like families. Little kids get disciplined in gangs. When a little kid drifts into a gang, he doesn't just get a gun thrust into his hands. He's gonna get homeboy love, which is pretty potent."[35] One such young man in 1967 returned from an extended period of incarceration and was to be seen with his homeboys every hour of the day, as if he had no family, no other place to be. He was always with the gang, and it soon cost him his life. A rival group raided his gang's territory, and of course there he was with the others. He tried to run off by himself but climbed an outside set of stairs only to find the door at the top to be locked. As he turned, he was shot and killed instantly. He was not a particular target; he was simply more available, more likely to be there because belonging was so important. I had known him for just a week.

In addition to some individual deficits and aggressivity and the identity benefits of belonging to the gang, several other characteristics have consistently been noted by gang researchers. For instance, there is the undeniable excitement that attends the anticipation of gang activity. I stress the anticipation of gang activity because, as I've said before, the most common gang activity is inactivity. But I spent many hours watching gang members animatedly discussing events—past events, rumored events, proposed new events—and emotionally feeding off these much as they might reenact an Arnold Schwarzenegger or Clint Eastwood movie. They rehearsed and relived the battles, embellishing them with little concern for reality. Many other gang writers believe that this excitement over challenges and battles serves to reassert gang members' masculine self-image in a context in which personal success in other endeavors is less common.

In earlier decades, perhaps more than is the case now, another regular feature of gang life was territoriality. *Turf, barrio, set,* and *'hood* (for neighborhood) were terms for the group's life space. Although the analogy often made to animals' territory and "marking" by leaving glandular scents is clearly inadequate, the sense of having a territory and often "marking" it with written graffiti was common to most gangs and shared by most members. Thus beyond the group, there was something tangible to identify with and call one's own.

With this came territorial rights and the obligation for territorial defense. One reason for joining the gang then became the defense of oneself and one's kind. Since other gangs had the same rights and obligations, one needed protection when the lines were crossed. Youth concerned for their own safety, in their home areas or immediately outside, could view the existing gang as a source of potential protection. Joining, for some, was seen as self-protection. Several young area boys complained to a notorious gang leader that they were being pressured to join groups that they didn't want to join. He told them to say next time that he was their cousin and he didn't like this recruitment. "Do you know," he told me, "I ended up with a whole schoolful of cousins!" In most cases, the need for protection is not really present or is exaggerated. Most inner-city neighborhood youths do not become (or have to become) gang members. But the perception of threat in a gang area is real, and some youths are more affected by it than others. If they join for protection, they might very well have joined for other reasons, in any case.

Finally is the issue of material rewards. Gang members commit crimes, many of which are property crimes. It's hard to make a living as a gang member, but it's not hard to make a score. Whether through store theft, auto parts burglary, strong-arming, or occasional drug sales, gang members have always found ways to get pocket money. The lesson was not lost on potential recruits. Even though most members did not join in order to commit crimes, the excitement and loot that accompanies crime are certainly added incentives for many. Nor is it the case that recruits were criminal virgins before joining, but joining would certainly be seen by many as facilitating illegal pursuits.

I've been dealing here with individual characteristics of gang joiners. I have not stressed in this context their ethnicity, socioeconomic status, neighborhood characteristics, and the like. These more structural variables will be a focus in Chapter 7. The point here is to consider what it may be that has led some youths in a gang area to become members when most have not. Another way of posing the question is to ask what kinds of variables we might select to study about people if we wanted to predict, before they were recruited, which youths would and would not be likely to join gangs. As a final comment on joining then, I would recommend the following as a minimal list:

- Low self-concept.
- Admitted involvement in violence.
- Defiance of parents.
- Deficits in adult contacts.
- Social disabilities or deficits.
- Deficient school performance, both academic and disciplinary.
- Limited repertoire of skills and interests.
- Poor impulse control.
- Early conduct disorder.
- Early onset of delinquency.
- Perceptions of barriers to jobs and other opportunities.

For each of these, I could cite from one to a dozen references that point to it as a predictor of gang involvement. But note that this is not a list of very positive traits. Despite the cameraderie and bravado that mark some gang gatherings, the typical street gang has been, at best, only a partially satisfying arena. Jim Short described the situation well:

> Gangs seem to promise more than they can deliver.... You join the gang for protection, and yet being in the gang makes you more vulnerable. You join the gang for excitement, and you spend most of your time waiting around for something to happen.... You join the gang for a sense of belonging, and yet the gang has proven to be one of the most undependable forms of human association.... You join the gang for status, and yet that exposes you to the status threats that all youngsters experience, but without ... supportive adult(s).[36]

Cohesiveness Revisited

The degree to which gang members were depicted in the 1950s and 1960s as needing their gang affiliation is tied directly to gang cohesiveness. The achievement of a group identity, the claim to social status that attends gang membership, and the sense of belongingness that can even yield the "gang as family" commitment are feeders of gang cohesiveness. Although this cohesiveness is more than the sum of its individual parts, it is most certainly a function of them.

But because so many individual needs are played out in the gang, there tends to be a rather low focus on group goals as such. Gangs are not committees, ball teams, task forces, production teams, or research teams. The members are drawn to one another to fulfill individual needs, many shared and some conflicting; they do not gather to achieve a common, agreed-upon end. Thus gang cohesiveness is rather tentative. LaMar Empey summarized the writings on gang cohesiveness in the 1950s and 1960s:

> The picture that is painted suggests that gang members, like inmates in a prison, are held together, not by feelings of loyalty and solidarity, but by forces much less attractive. It is not that structure is lacking but that it is defensive and highly stylized, not supportive. Group members stay together simply because they feel they have more to lose than to gain by any breach in their solidarity. While they may appear to the outsider to be dogmatic, rigid, and unyielding in their loyalty to each other, the sources of this loyalty are not internal but external. Remove the pressure and you remove the cohesion.[37]

Lewis Yablonsky was one of the first writers to relate gang cohesiveness to gang intervention programs. He saw group cohesiveness as so low, the structure so fragmentary, that he worried about those who would "project group structure onto gangs." He concluded:

> Approaching the gang as a group, when it is not, tends to project onto it a structure which formerly did not exist. The gang worker's usual set of notions about violent gangs as groups includes some of the following distortions: (1) the gang has a measurable number of members, (2) membership is defined, (3) the role of members is specified, (4) there is a consensus of understood gang norms among gang members, and (5) gang leadership is clear and entails a flow of authority and direction of action.
>
> These expectations often result in a group-fulfilling prophecy. A group may form as a consequence of the gang worker's view.[38]

Significantly, Yablonsky saw the limits on gang cohesiveness as deriving from the internal conflicts of the members themselves; the constituents of the group (or "near-group," as he preferred) provided the constraints on its groupness. This is akin to our conclusion that gang cohesiveness derives more from external than from internal sources and is close therefore to Empey's characterization of the consensus of the literature as quoted earlier. But equally important is that this tentative cohesiveness contained the seeds for further increases if the external sources were brought to bear. These sources could be well-meaning detached worker programs. They could also be suppressive law enforcement programs. Chapters 5 and 6 on handling the gang problem will thus rest on the foundation established here in Chapter 3.

For example, modern-day rap groups and break-dancing groups and "taggers" (youth who concentrate on self-enhancing graffiti) are not, in my estimation, street gangs. Yet there have been a number of instances across the country of such groups getting into contests with one another. In these contests, they represented their own neighborhoods or communities. The contests, in the cited instances, turned into genuine rivalries with territorial overtones and thence into rival gang situations. If not alerted to such a possibility, the best-meaning sponsor of such innocent youth culture activities can inadvertently set the stage for gang development.

Does this seem far-fetched? It does to me, yet it has happened. Good intentions don't guarantee good outcomes. I've seen numerous instances of adult-sponsored truce meetings between gangs that had a boomerang effect. They reinforced the rivalries, solidified the leadership of the selected truce spokesmen, and increased the cohesiveness of the warring factions. And this was a common occurrence, not far-fetched at all.[39]

The Positive Functions of Gangs

This matter of using the existing gang structure to meet positive ends, as in truce meetings, brings me to a most difficult discussion. In a nutshell, the issue is this: Traditional liberal thought about gangs says that they are the result of a mismanaged society, that they have untapped talents and resources, and that they can be turned to good causes by good people. Gang writers for decades have worked around the edges of this issue. It was a fashionable concern in the 1960s because of the existence of gangs in the very areas that involved the Community Action Programs during the Kennedy/Johnson War on Poverty era.[40]

There were four parts to the gang-as-positive-mechanism philosophy. I'll comment briefly on each.

Gangs facilitate their members' self-actualization. Yablonsky captured this point in his quotation from a gang work–training manual:

> Participation in a street gang or club, like participation in any natural group, is a part of the growing-up process of adolescence. Such primary group associations possess potentialities for positive growth and development. Through such a group, the individual can gain security and develop positive ways of living with other individuals. Within the structure of his group the individual can develop such characteristics as loyalty, leadership, and community responsibility.[41]

As long as such a statement is tentatively offered—"possess potentialities," "can gain security," "can develop"—it is acceptable. There are many youths who do benefit from their gang affiliations. But there also are many who suffer the consequences of membership. Aside from the personal harm of being handled by the justice system, they also are victimized by other gang members; they are more readily cut off from educational resources and the public reward of high school completion; they gain less access to job skills and opportunities; in prolonging their peer relations, they are exposed to fewer positive adult role models and trainers and thus develop fewer useful social skills; the list goes on, but the point is clear. The gang is not like "any natural group," as the manual's phrase puts it. As a unit effectively cut off from the connections enjoyed or fostered by other natural groups, such as the scouts, the French club, the church youth group, and the local pickup basketball team, the gang deprives its members, already self-selected by some deficits, of too many social advantages to counterbalance its positive offerings.

Gangs provide local empowerment. During the 1960s, giving more decision-making power through the use of War-on-Poverty funds rightly became a feature of various community action programs. The principle was extended in a number of cases to gangs as well. Most infamously in Chicago, major street gang organizations became recipients of public and private grants to do community organization, develop social welfare programs, and enter into profit-making commercial enterprises. The Devil's Disciples and Blackstone Rangers reportedly received almost a million dollars in federal poverty funds and a $50,000 private grant for legal defense programs. The rival Vice Lords were reportedly given at least six public and private grants for various projects. A Senate investigation and other reviews uncovered an astounding picture of the consequent corruption, fraud, and crime that resulted from these grants.[42]

More recently the Blackstone Rangers, recast first as the Black P. Stone Nation and then as the El Rukns, were indicted for multiple crimes, including contract murders and conspiracy to supply arms to Mu'ammar Gadhafi's Libyan forces. From little acorns great oaks are grown. There is nothing in the "natural" structure of a street gang that is appropriate preparation for community service or corporate success on a legal basis. Gangs usually don't even make good illegal commercial organizations, as is evidenced by their usual ineffectiveness in the crack distribution business. However, such experiences soon leave the well-meaning social activist's mind.

Gang members' skills can be coopted for social goals. I watched with considerable interest an attempt to turn several Latins from active gangbanging to active involvement in La Raza, the Chicano-based social power movement of the 1960s. The attempt was being made by a church leader trained in the social change model of Chicago's famous Saul Alinsky. The targets were several core Latin members who were hanging around the church's youth center. The trouble was that they were at the center early in the day and were cruising rival gang areas later in the day. The gangbanging hadn't stopped; it had found a cover, and there was no genuine coopting of the core members' energies into the social movement.

In the early 1960s, a respected gang leader, Hamilton, had been in secure confinement for almost a year, and his return was excitedly anticipated by the Operators. But Hamilton returned with a veneer of reform, talking about the group's devoting its energies to the Congress on Racial Equality (CORE), the black activist organization that provided much of the grassroots leadership for civil rights in that era. When Hamilton finally presented his proposal formally at a scheduled gang meeting, not only were his ideas rejected, so was he. For a while he risked being badly beaten in the session, but he read the situation well enough to retire before that happened.

The group saw little of Hamilton in the months that followed. His former high-status reputation collapsed as he switched his loyalty from gang to cause. He was left there, untutored by CORE in using his assets with his former close pals. The Operators had their own problems—personal problems, gang problems—and had little capacity for engagement in more abstract causes and movements. In this, they were no different from gang members across the country. In another personal example, a young member of the Generals whom I got to know joined the politically active U.S. organization in the 1960s but became involved in the murder of two members of the Black Panthers, a rival black power group. He was caught, convicted, escaped, and recently turned himself in to the U.S. embassy in Surinam in 1994, after almost thirty years of his life had been totally wasted. It's not that gang leaders didn't have skills. The problem was with the simplistic assumption by civil rights activists that those skills could be transformed into civil rights organizing and leadership activities. Leadership is not a homogeneous phenomenon.[43]

I should quickly add that some instances of transformation from gang to social cause were successful. Chicago, the most politicized of all gang cities, provided both positive and negative examples. But the positives were the exceptions. Street gangs are no better structured for civil rights engagement than they are for organized drug distribution, but it's the exceptions that continue to capture the public imagination.

Gangs serve to stabilize disorganized communities. In the midst of an era of urban riots, one writer claimed, "urban youth gangs, particularly those which are highly aggressive, are a stabilizing influence in a community characterized by severe lack of social and economic opportunities."[44] That any organization is better than none seems to be the point. This thesis is hard to accept. The mere existence of gangs can hardly be stabilizing, even at their best: "Delinquent gangs were rather 'innocent' participants in the broad social trends of the times."[45]

At another extreme, it has been claimed that gangs have developed mutually supportive relationships with community institutions, a symbiosis that includes gang members' genuine commitment to the needs of local community residents.[46] But this seems more ideological than historical. For the most part, street gangs have not been sufficiently entwined in the institutional structures of the community to have much effect on community stability.

But more to the point of all four of these theses regarding the positive functions of gangs—for their members or their communities—is that the typical street gangs aren't sufficiently stable themselves, internally, to be particularly helpful instruments. What a youth gets out of a gang that is positive for him (or her) is more related to the individual needs brought to the gang. Some of these, such as the need for identity, for belonging, and for protection, can indeed be served by membership. However, even this gain is too often counterbalanced by the damages of gang membership.

To expect these fragile organizations to become bulwarks of community activism is, far more often than not, simply unrealistic. It is an expectation based less on the nature of the gang than on the hopes and values of the activist. Thus gangs in community action are more often used than empowered. In the project assessing effectiveness of gang workers with four black gangs in Los Angeles in the early 1960s, one death occurred over a four-year period, an accidental stabbing during a high school rally. More deaths came a bit later, when a handful of our eight hundred gang members became entangled in an internecine struggle between the Black Panthers and rival black activists. The setting was a meeting at UCLA that turned into a shoot-out. This was not empowerment, except in the sickest sense.

Street gangs are their own call for social intervention. We don't need to respond to our other social values by taking advantage of the gang presence or by using it without truly understanding it. The failures in using gangs as positive forces in past decades constituted an important lesson. As we shall see in Chapter 5, it was a lesson soon forgotten, as the attempts continue today.

4

The Contemporary Street Gang Situation

Once disorganized and lacking both the leadership and resources, the gangs of today are organized, are goal minded, and well equipped with weapons, vehicles, electronic gadgetry, and fat budgets. The gangs of today, whether they realize it or not, are becoming business majors. They are involved in merchandising, franchising, retail and wholesale sales, and market expansion. Their product is illicit drugs, and the rock cocaine is the flagship of their line.

Testimony from Lieutenant Larry Carter, Inglewood Police Department, to the California State Task Force on Gangs and Drugs (Final Report, California Council on Criminal Justice, Sacramento, California, January 1989)

What's wrong with this dramatic testimony, and so many other recent statements from law enforcement officials across the country, is that it reflects the limited view of those primarily experienced in drug markets, not street gang realities. It projects a conspiracy-oriented mentality that goes well beyond the capacity of most street gangs. It ignores the vast number of street gangs and gang members across the nation who are only tangentially involved in drug sales, being more occupied—as they have been for fifty or more years—in the more mundane problems of daily living, minor crimes, and the prospect of serious violence.

Much of what I have to say in this chapter about the contemporary street gang situation stands in opposition to the portrayals offered not only by some law enforcement agencies but also by many politicians unacquainted with the data on modern street gangs. Issues of gang definitions, gang proliferation, the nature of street gang structure and crime, and, once again, the gang–drug connection require careful review before we consider public policies to deal with gangs.

Prevalence

Denial

The advent of street gangs to a community is not a happy occurrence. It suggests that the community has failed in its task of socializing some portions of its youth. Since street gangs are composed mainly of minority youngsters, the failure also implies the specter of institutional racism. In addition, the continuation of these gangs, now comprising a gang problem, reflects on the inability of community institutions to control or roll back the problem. Social agencies, schools, and the justice system appear impotent and thus face a public relations problem on top of their gang problem.

It is no wonder, then, that communities or segments of them often deny the existence of street gangs in the face of contrary evidence. I recall the comments of a western police official in a high command position some years ago who described his city's earlier denial of gang problems. As I remember his story, an adolescent girl from a prominent family was brutally assaulted by purported gang members in this "gang-free" city. The hue and cry went up on a Friday, and the following Monday the police department reported to the city council the confirmed existence of five large street gangs and complete rosters of their memberships. Now either that was one incredible weekend of intensive police intelligence operations or the police had been tracking and recording gang data for some years while simultaneously denying the presence of gangs. My police official was quite clear on this point. The weekend activity consisted only of writing up the report for the city council; the gangs and the members were already well (but secretly) documented.

So another of our cities moved from denial to admission. In the police survey I described in Chapter 3, approximately 40 percent of the respondents commented that their departments' initial response to the emergence of gangs was to deny their emergence. The responding officer was asked, "When your department first became aware of street gangs, how did it respond—was there denial, or someone assigned to check them out, or a unit developed, or what? Can you describe what the response was?"

There were some "don't knows," of course, from respondents who weren't around or involved at the time of gang onset. Thus a 40 percent denial is probably an undercount. Most denials were said to be for public relations purposes, but a few were related to gang cohesiveness; that is, admitting to the presence of gangs would legitimate them and thence lead to their becoming more cohesive. "For fear of coalescing the gangs," as one respondent properly put it. Another, the chief in a small city, admitted having "youth groups" with drive-bys and homicides but added, "We got all city officials three years ago to reject gang terminology. When we do have these shootings, we call them deliberate murders, not drive-bys, which would trivialize them." The public relations strategy was captured by the officer who noted, "We're a tourist city," and by another who described it as "institutional denial; the city council, the mayor, and the department are still doing it." This latter city acquired gangs in 1987. Others reporting denial as still taking place listed their gang onset years as 1989, 1988, 1985, 1983, and the early 1970s; one was in the 1950s and another "before my time." Denial on an official level can last quite a while, obviously. One respondent characterized it as "very heavy denial," and another as, "denial, big time."

A few cities reported to me in 1991 that they were free of gangs, but less than two years later reported that they had street gangs dating back generally from one to five years (and in one case, back to the 1960s). There were seven such cities among the fifty-five in my survey that reported being gang free.

Given such widespread denial in the initial stages and the vehemence of it in some cases, it is possible that some of the 55 officers from my total of 316 interviews (261 reporting the existence of gangs) were incorrectly denying at the time. I can, in fact, cite five such cases.

1. An official of a large southern city claimed no gangs but soon thereafter sent me a letter specifying that the city was now acknowledging a street gang problem.

2. An official of another good-sized southern city denied having gangs, but the city was portrayed by police gang experts elsewhere in the state as clearly being a gang city. That city's county reportedly had a heavy gang problem, and this city was the only large one in the county. During a reinterview in 1992, the city officials admitted to the gangs' onset in 1990, with ten gangs and two hundred members.

3. An official of a southwestern city responded, very deliberately, that it had "a policy not to have gangs." Other cities had specifically nominated this one as having gangs, but the respondent refused to deviate from his chief's policy concerning gang denials. Soon after the survey, there was a change in chiefs, and this city suddenly had street gangs.

4. A large eastern city had had a street gang problem in the 1960s but then became known as an anomalous gang-free city. More recently, it became one of the very worst drug sales and violence locations in the nation. The respondent, a very high ranking official, asserted that the city had no known street gangs. When information came to me that threw serious doubts on this response, I called the

official back and was told that the department was now investigating the issue to determine what in fact was going on. Soon thereafter, the chief announced that the drug problem was being fueled by "youth gangs" with neighborhood-derived names, yet a 1992 detailed homicide report from this city makes absolutely no mention of gang connections.

5. My last example is Pittsburgh, which I name outright because the folks there were so open with me about their own confusion as to whether or not the city had gangs. My initial request for an appropriate respondent to be interviewed brought the chief. I had already learned that when a chief wanted to handle the interview, it usually meant a no-gangs response would be forthcoming. And so it was. The chief reported no gang problems; probably, he speculated, because the city's minority population—mostly black—was widely dispersed rather than concentrated.

But the Pittsburgh plot slowly thickened. A large, longitudinal research project on youth crime was being carried out there, similar to those in Denver and Rochester and, like those other two (see Chapter 3) included at my suggestion some questions about gangs in the youth interviews. When I asked one of the two senior researchers about the results, he told me that there were no gangs reported in Pittsburgh. His partner immediately countered, "Oh yes, there are."

The researchers kindly sent me some of their data to let me resolve the issue. Unfortunately, the data don't help much. The responses, coming from young interviewees in high-risk areas of the city, do indeed report gang membership, but to groups whose descriptions can only be called somewhere between superficial and whimsical. These youngsters seemed to be "playing" at being gang affiliated—school-based "wannabes" rather than committed gang members.[1]

In fact, Pittsburgh does indeed have compacted, inner-city black areas. On this, the chief was mistaken. It also has a few very visible, active drug market areas—a lot of crack cocaine—and a number of former crack houses. There are drug distribution gangs in Pittsburgh, and some heavy importation from outside. Housing Authority cops are particularly upset about the situation. And yes, there are some clearly wannabe groups in selected inner-city schools. But neither the drug groups that evolved independently nor the wannabes who are evolving out of the national diffusion of gang culture can be fairly described as street gangs, as I've used the term in this book. Not yet, at least, but I wouldn't bet on this being a stable situation.

Police often explain their suspicions about a likely offender by referring to the "duck metaphor." They say, "If he walks like a duck and talks like a duck and acts like a duck, chances are he is a duck." Well, thanks to the diffusion of gang culture in America, through films, the press, news shows, MTV, gang seminars, and the like,

American youth are learning what it is to be a "duck." They learn gang names (Crips, Bloods, Latin Kings, Vice Lords, and so on), gang colors and symbols, gang attitudes and behaviors. In short, they learn to walk gang, talk gang, and act gang. It should not surprise us, then, if they become gang. All they need is to develop some sustainable rivalries: my school versus yours, my housing project against yours, my neighborhood against yours. It's happened in other cities; watch out Pittsburgh, the young ducks are in the pond.

It has become fashionable, by the way, to suggest that police will finally admit to gangs, to accept the notion of a gang problem, only when a particularly heinous event takes place. In Denver, it was the assault on the white daughter of a prominent family. In Columbus, Ohio, it was the independent assaults on the mayor's son and the governor's daughter. In Joliet, Illinois, it was the killing of a police officer. But in fact, these are the exceptions. Denial never seems to solve the problem; it only prevents the mobilization of resources. Eventually, the problem is faced and, sadly enough, faced out of necessity in city after city after city. As one of my students noted in class, "They bury their heads in the sand, and then they get bit in the butt."

The New Numbers

We should have known that gangs would proliferate, I suppose. The indications were in the criminological literature, if not in the public domain, in various works.[2] True, I was informally collecting my own list of cities reporting gangs, some coming from our California research, some from research reports and colleagues, some from media reports, and some from communities seeking consultation.[3]

I had a list in 1990 of just over 50 cities, each having been corroborated in some way. By January 1992 the number was more than 100. And as the reader knows, my police survey and other work have now advanced the number from the 261 in the survey to far more, 800 or more towns and cities, as documented in Chapter 2. The number is stupefying. It even includes 91 towns with a population of less than 10,000. I'll have more to say about the reasons for this situation in Chapter 6, but meanwhile we should look at the rate of growth. It has been anything but steady. The dates of onset in these cities are summarized in the following maps, showing the accumulated pattern pictorially that is also summarized first in Figure 4-1 on page 91, based on those departments that could report the date of onset with some certainty.

I urge the reader to look over Figure 4-1 to get a visual feeling for what been happening. It amounts to a seriously accelerating problem: increases of 74 percent by 1970, 83 percent by 1980, and 345 percent by 1992. There is no support here for those who believed that gangs were quiescent during the interregnum between the classic

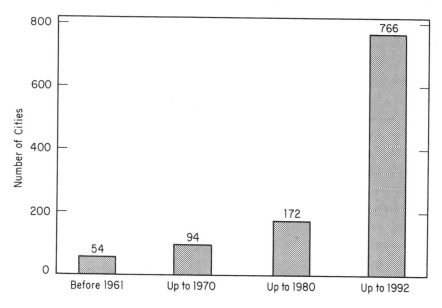

Figure 4-1. Growth in Gang-Involved Cities (N = 766).

gang research period of the 1950s and 1960s and the renewed interest in response to an explosion in the 1980s. "Youth gangs aren't 'back.' They never went away, except in the media," noted Walter Miller.[4]

A 1992 map from a state department of justice, based on "data" supplied from the Federal Bureau of Alcohol, Tobacco, and Firearms (ATF), identified 33 states and 123 cities with Crips and Bloods involved in assaults, burglaries, homicides, narcotics, and robberies. Leaving aside the obvious point that the ATF has no research capacity for deriving or confirming such an extensive Los Angeles invasion of the nation, the implication, like that of the Drug Enforcement Administration's (DEA) interstate highway model of gang migration, is that the Crip and Blood crack franchising is the cause of this 123-city takeover. This is typical of the sort of information offered to the public and used to help formulate public policy. Yet the emergence of gangs in many of the gang cities listed actually predates the emergence of crack in the mid- and late 1980s. Crack clearly exacerbated the problem with some gangs in some cities (but until recently, only those cities with black street gangs), but few cities needed crack to initiate street gangs. Gangs were already there, and so crack was an add-on, not a cause.

Figures 4-2 to 4-5 on pages 92–95 show the cities with gangs through 1960 and each successive decade. California, and Southern California in particular, was initially the most affected area. The big cities of the East revealed a contained pattern rather than a set of regional proliferations. But through the decades after 1960, concen-

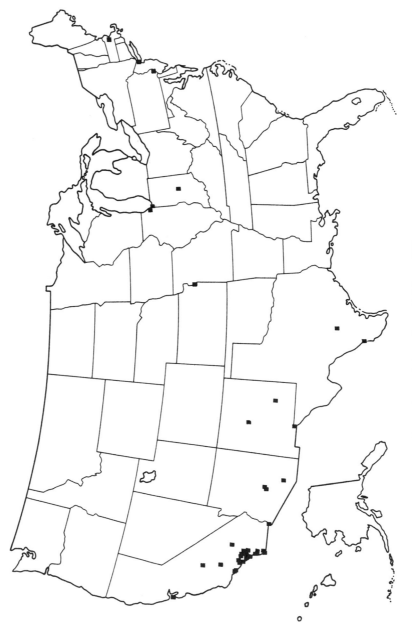

Figure 4-2. Cities with Gangs Through 1960 (number of cities = 58).

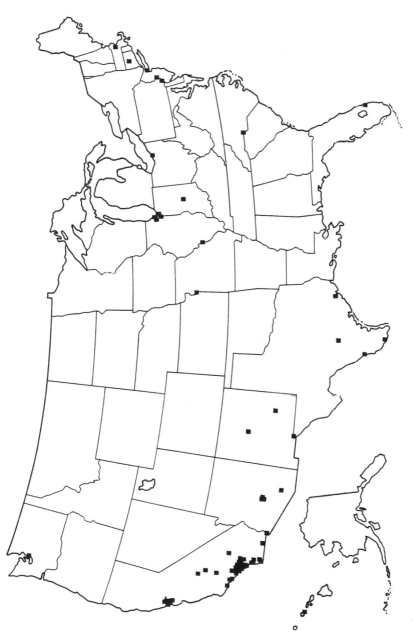

Figure 4-3. Cities with Gangs Through 1970 (number of cities = 101).

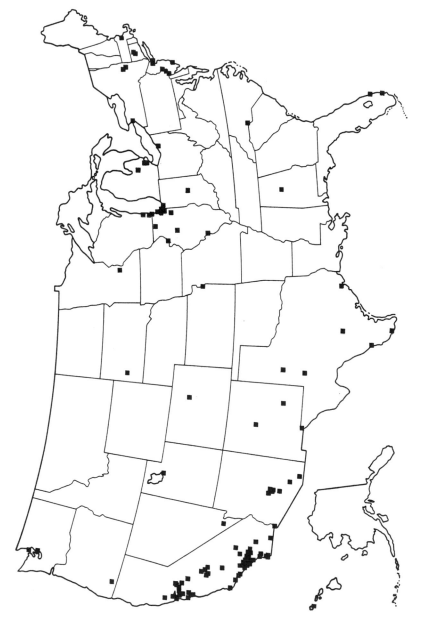

Figure 4-4. Cities with Gangs Through 1980 (number of cities = 179).

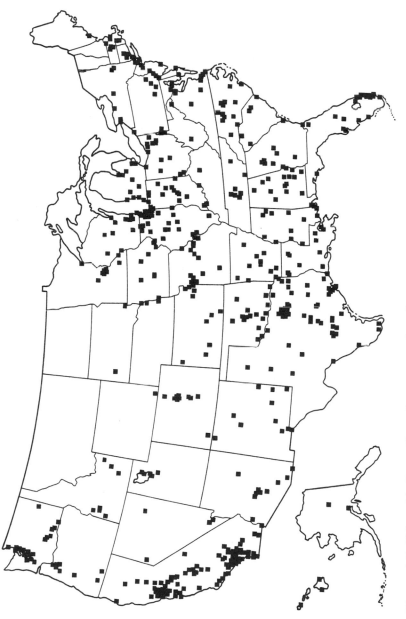

Figure 4-5. Cities with Gangs Through 1992 (number of cities = 769).

trations began to appear, notably in the Southwest, in the industrial Midwest, and then in the South generally. Although federal enforcement agencies have reported the proliferation as a spread from Los Angeles outward and occasionally from Chicago and Miami, our interviews and the results of the Maxson–Klein national gang migration study point to a very different and far more discouraging picture. Only 5 percent of the gang-involved cities were aware of any migration of gang members into their jurisdictions before the arrival of gangs in those cities. Thus the proliferation of gangs is not a drug market-driven phenomenon, although the crack trade has been a parallel development. The level of gang migration from Los Angeles cannot begin to account for the appearance of up to eight hundred gang-involved cities and towns.

Rather, we have close to eight hundred homegrown problems. Some of them are exacerbated by regional patterns of migration, but most of them are parallel, independent, and sadly understandable responses to fundamental changes in American social ecology. As an example: Of 267 cities reporting gangs by 1985, only 87 reported any migration of gang members from other cities.

Figure 4-6 on page 98 shows the number of gang cities of each population size. Gangs are no longer a big-city problem. The accelerated emergence of gangs, I will suggest in Chapter 7, is a function of a growing urban underclass and the widespread diffusion of gang culture through the media and other sources of dissemination. Big cities are by no means the only recipients of the urban underclass or of gang culture diffusion. Rather, these are to be found across a myriad of locations—thousands of towns and cities.

Figure 4-2 should be interpreted with some care, since the four categories of city size are not equal. The 178 cities of 100,000 or more population that have gangs include almost all cities of such size. The 179 gang cities in the next category, the 344 cities of the third category, and the 91 towns of the smallest category obviously comprise smaller proportions of the total numbers of cities of those sizes. Yet it is clear that the gang proliferation, if it continues, will necessarily do so increasingly in our smaller cities—the large ones have already felt the effects disproportionately. It is clear also that our best prevention and enforcement efforts have totally failed to stem the tide: The cities that are former gang cities can be counted with fewer than the fingers on one hand.

Nonetheless, there is time to mount counterefforts. Remember from Chapter 2 that the proliferation is composed principally of cities with relatively few gangs (fewer than ten) and relatively few gang members (fewer than five hundred). Communities that get serious about preventing or reducing gangs can still do so. State and federal agencies that want to contribute substantially to this process can still do so, as well, but not by frontal attacks on gangs alone. Gangs are by-products of their contexts, and it is these contexts that must become our principal target.

Certainly, the approaches to gang control, reduction, and prevention must differ from small cities to large, for example, from Tyler, Texas, to Houston; from Durango, Colorado, to Denver; from Beloit, Wisconsin, to Milwaukee. Isolated and regionalized gang situations undoubtedly raise some different questions as well. Just to illustrate the complexities that arise from the nature of the gang-city dispersion, let's consider the maps again, which result from combining two sources of information, our own interviews and questionnaires, described in earlier sections, and a geographic information system file for the entire nation.[5]

Figure 4-2 shows that the early gang years, despite public images of gangs as eastern-city problems, were principally a southwestern phenomenon. Southern California, in particular, experienced street gangs in the years before 1960. Some minor proliferations took place between 1961 and 1970, with a few eastern cities added to an increasing number in California in particular (see Figure 4-3).

By 1980, as seen in Figure 4-4, both Northern and Southern California showed large clusters of gang-involved cities. Wider-spread clusters now appear along the northeastern population corridor and in the Midwest, centered in Chicago. The southwestern states begin to become pockmarked, whereas the Plains states and the South generally remain gang free.

Figure 4-5 is obviously the most striking. The gang-city proliferation accelerated after 1980. The California picture grows; the East Coast corridor is now solid; the Midwest has filled in remarkably, and what earlier looked like a southwestern expansion has clearly become generally a southern pattern. Indeed, only four states have no gang-involved cities.

I was particularly struck by a recent report of gang activity in South Lake Tahoe, the resort area on the Nevada–California mountain border. I drove through there during a visit to Reno to lecture on gang proliferation. My drive to Lake Tahoe was supposed to get me away from gangs and into fresh, forested air removed from urban ills. But the motel and casino strip of South Lake Tahoe has drawn a large minority of Hispanics and Filipino immigrants to fill the low-paying jobs in this town of 21,000 people, enough to yield the first gangs, named South Side 13 and BNG.

Actually, the Tahoes of the nation are in some ways our most intriguing gang cities. Of the gang cities of less than 10,000 in population, 94 percent acquired gangs after 1984. Of those between 10,000 and 50,000, the figure is 78 percent; of those between 50,000 and 100,000, the figure is 64 percent; and of those with more than 100,000 in population, the figure is 56 percent. These data, taken from the Maxson–Klein migration study, clearly show that the proliferation of gangs is increasingly affecting the more numerous small cities. It should be noted, however, that the smaller cities, those with less than 100,000 in population, yield expectedly smaller numbers of street gangs. Among the gang cities reporting these data to us, 62 percent of the smaller cities reported five or fewer gangs, as compared with 26 percent of the larger cities. Twenty-five or more gangs are reported in only 3 percent of the smaller cities, but in 32 percent of those with more than 100,000 in population. Judging from the results of the gang migration study, the total number of gangs and gang members in the United States in 1991, recognizing the problems of definition and incomplete data, stands at more than 9,000 gangs and at least 400,000 gang members in any given year. Gang member turnover during a three- or four-year period would double the latter figure for such a period.

At the same time, in some of the larger metropolitan areas, there is not likely to be one characteristic gang situation. Look at the Los Angeles area on the maps; dozens of cities are involved, varying as much in their makeup as does almost any equal-numbered set of cities drawn from across the maps. Another case in point is St. Louis. Communities in the metropolitan area of that city are differentiated as being most settled, least settled, and intermediate. Accordingly, the numbers, characteristics, and stability of gangs there differ with the degree of community stability.[6] Variation both within and across cities

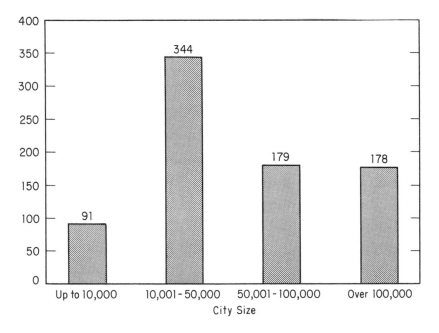

Figure 4-6. Gang Cities (792) by Population City Size.

is common and must be kept in mind as we discuss the American street gang.

In talking with people in a number of the newer gang cities, I have frequently gotten the impression of the sorts of milder street gangs commonly found in the 1950s and 1960s. This seems to be true of the Honolulu gangs and is nicely depicted for the emerging gang situation in St. Louis by Scott Decker and his colleagues.[7] For instance, a film crew spent many days in 1993 in the city of Thousand Oaks, California, and then sat down with us to discuss their observations. Thousand Oaks is a suburban community in the golden, rural hills on the northern fringes of Los Angeles, an upper-middle-class jurisdiction that has blossomed into a population of more than 100,000.

Five ganglike groups are said to have emerged since the late 1980s, one territorial Hispanic gang and four less territorial white groups with some black and Hispanic members. All are named groups totaling about 150 members. The first shooting reported to the police was in 1989, and twenty-five "serious" violent incidents have been reported over the past five years.

Weapons are old style: bats, chains, and the like, with very few firearms. "A shooting sends them into hiding," reported the film crew. Drug sales are low and organized crime is light. Two local tagger crews (see Chapter 7) are also in evidence. Yet despite the borderline "ganginess" of these groups, they use hand signals; wear the expected baggy pants, hightop shoes, and Raider caps; use gang argot; and have adopted Hispanic accents in their spoken language (as taken from the prison gang film *American Me*, which they view over and over).

These groups illustrate the diffusion of gang culture (to be discussed further in Chapter 6) which in turn produces gang behavior. These are not classic 1990s street gangs, not yet, but they are now engaging in "where are you from" drive-by challenges, one of which was witnessed by the startled film crew. Less than a month after the film crew's description, news reports appeared of a confrontation between about thirty members of two of the Thousand Oaks gangs at a local park. Four female members were beaten with bats and other implements after their male companions fled from the scene.

Cities on the Cusp

If gang cities emerge at such a frightening rate, then it stands to reason that a number of them (like Thousand Oaks) not yet properly listed on the gang side of the ledger may soon be so listed. Indeed, some of my preceding examples were "on the cusp." Pittsburgh is a dramatic example because of its size, its acknowledged drug gangs, and the appearance of the copycats or wannabes in the longitudinal youth survey. Other examples emerged in my survey of police informants.

Some of them give the sense of a city or town that is close to street gang status or might be seen as such from outside.

- A southern university town that saw groups start up in 1989 with gang names and symbols and then die down. The same up-and-down pattern occurred in 1991.
- A northwestern city that reported four to six "youth groups," averaging 15 years of age, locally structured—"not gangs in the Los Angeles sense"—with gang names, graffiti, and some light drug use. They are not territorial, with little rivalry, no drive-bys, but a variety of crimes. They are 97 percent white kids.
- A small, midwestern city with one gang of fifteen members in a housing project. It is poorly organized but has a two-year history. Can there be a one-gang city? The police official interviewed there believes he has one.
- A small, central California town with two groups that started in 1991, now grown from twenty to forty members. They are school based, exhibiting no weapons, no assaults, and no rivalries but have engaged in graffiti writing, burglaries, and minor fights. "Our kids are being taught by outsiders."
- A southwestern town that originally acknowledged local street gangs in a survey conducted by the state attorney's department. In my (later) interview, the assistant chief told me that the department had overstated its situation. It did have copycat groups inspired by the gang movie *Colors*. These are school-based wannabes, poorly organized, with graffiti and minor fights but not seriously criminal. No drive-bys, firearms, or turf mentality, said the assistant chief. "We all got excited about it, set up alternative activities, and sent two detectives off for special gang training."
- Another southwestern location reported twenty-seven documented little groups, either school or neighborhood based. They are mostly juvenile copycat types with graffiti and a few drive-bys. Busing practices limit their after-school coalescing. "They're not gangs, not yet," said the respondent.
- A Texas city reported three drug distribution gangs, small with a few local recruits, originating in and dealing crack out of Los Angeles. There is also a street group in a housing project, about a dozen kids without a rivalry, although the next project is trying to develop a counterpart.
- A southeastern city in which six groups, starting in 1969, recruited about forty members in all. These are strictly drug dealers, starting with crack from New York City distributors. They are limited to housing project areas.

There are several fairly obvious dimensions running through these descriptions. These include the seriousness of the criminal involvement, the self-acceptance of the gang label, and descriptors such as territoriality and rivalries, migration effects, cultural diffusion of gang symbols, and the distinction between street gangs and drug gangs. Thus coverage of the contemporary gang structure and function is critical to the proliferation issue, and I'll provide more of this later in the chapter. Here, I want to emphasize again the reality that what is or is not a street gang, and therefore what is or is not a gang-involved city, retains some element of arbitrariness no matter how carefully one tries to clarify "gang-ness." Of the eight cities "on the cusp," how many would you list as gang cities? None, perhaps, but wouldn't it be interesting to return to them in two or three years to see where they stand?

Structure

I don't need to go through the litany of structural characteristics that were covered in Chapter 3 concerning gang history. A sense of changes in street gangs into the 1990s, in addition to their proliferation, can be obtained by reviewing a selected few characteristics. I've chosen four of them: structures, ages, ethnicity, and female involvement.

Gang Forms

Joan Moore notes for one of her observed traditional gangs what is probably true for most:

> The White Fence gang of the 1980s is not the same as the White Fence gang of 1960: there are continuities, but there are also changes. What anybody "knows" about a gang in any given year—even a gang member's knowledge—may in certain specifics be out of date the very next year.[8]

We know that the traditional gang form—the cluster of age groups and cliques described in Figure 3-1 in the last chapter—is still to be found across the United States. They have been described in East Los Angeles, in Chicago, in Milwaukee, and in Columbus and Cleveland.[9]

I attempted to get at this issue in my police survey by asking the 261 respondents about the structures of the gangs in their cities. I offered some alternatives: single groups of members, traditional age-graded clusters, clusters of geographic "branches," cross-gang alliances. Many of the respondents were able to answer very usefully, but a number could not; a number of police gang experts knew little about the structural nature of their targets. Because they were concentrating on crime patterns and investigative issues, they found little relevance in the forms that their gangs took.

Nonetheless, the majority were able to respond, and with the following results:

1. The traditional gang still exists but is not the predominant form. With so many new gang cities, generational patterns have not yet developed (if, indeed, they eventually will). Thus the traditional gang form is more prevalent in the older gang cities and in the Southwest (including Southern California), where the Mexican-American "turf gang" pattern has long been transmitted across both generations and ethnicities. Sixty-two respondents reported traditional gangs.

2. In contrast, 171 cities reported the existence of single or autonomous gangs. This, then is the predominant pattern. Single, named groups, occupying smaller territories—perhaps school based, project based, block based—with shorter histories and fewer ties to neighborhood traditions than the traditional gangs have. From a gang control perspective, they should pose less serious problems, being smaller and less connected to other structures, and probably less cohesive. Suppression procedures (to be described in Chapter 6) might prove more successful with these autonomous groups, as long as they do not backfire by creating greater cohesiveness.

3. A third pattern, reported by fifty-three cities, was of geographically connected gangs. That is, rather than age-graded subgroups in the same territory or barrio, these were more like branches of the gang located in neighboring or separated areas, sharing affiliation but not residence. Certainly the Los Angeles Crips and Bloods follow this pattern to some extent, but the separations have become so large that Crip "branches" have become enemies. Their predecessors, the Slausons of the 1960s, were true branches of the same tree, with strong affinities among them.

I should add that many cities report two or three gang forms. Often this meant Hispanic traditional gangs and black autonomous gangs. The geographic branch pattern seemed to have a less ethnic tone. There also were ten mentions of "alliances," generally of the Chicago "supergang" variety described earlier and often called by the Chicago names—Latin Kings, Vice Lords, Disciples, and so on. Whether these are in fact more than affiliated branch formations I cannot say.

In reporting these new findings, I want to offer a caution to the reader, beyond that occasioned by police respondents who couldn't handle these structure questions. I also asked my officers about terminology applied to gang groups and members. The dismaying array of terms makes it clear that consensus is not a feature of the current American police knowledge of gangs, and it concerns me that the structural patterns reported may not be highly reliable. As a result, we are now initiating a national survey of gang structures.

When asked about how highly structured the gangs were, sixty-seven said loosely structured, and fourteen said highly structured. Most officers couldn't handle the question. The ratio is probably about right, but how striking it is that more officers couldn't respond.

What kinds of gangs? The diversity of terms is bewildering. Nine reported neighborhood gangs, twenty-three said turf gangs, twenty-four said "copycats," seven said "posses," three said minor criminal groups, and one said chapters. There also were twelve reports of school groups, four of "nations," six of cliques, two of taggers, one of crew, and ten of subsets or subgroups. The old 1950s–1960s term *sets*, common to black gangs in Los Angeles, received thirty mentions in the survey. Again, terminological consensus does not exist and probably understates the regularities of gang formation.

Terms for various levels within a gang formation show a similar variety, going far beyond the simple core–fringe topology. Now we get core, hard-core, and actives and rostered, confirmed, or certified (these latter reflecting the needs of new legislation). These are contrasted in the same gangs with fringe, associates, peripheral, marginals, and wannabes. What struck me in these conversations with the officers was that the terminology was far more varied than the substance to which it was being applied. Missing was the terminology of the gang members themselves—homeboys, brothers, OGs, *veteranos*, peewees, and so on.

It's hard to build a science on the base of data like these. Officials are applying self-satisfying terms to arbitrary distinctions. We badly need more careful research on gang structures, for at least two reasons. The first is simply to understand. The gang phenomenon is complex, and it challenges us as social scientists and interested laypersons.

Second, I am convinced that the success of attempts to control gangs and provide more useful lives for their members will differ depending on the types of gangs and gang members we confront. Treating all gangs as if they were simple variants on a homogeneous, monolithic pattern is silly, yet that's not an atypical approach these days. In particular, lumping together street gangs and drug gangs, as did the lieutenant quoted at the start of this chapter, is currently the worst error we make.

Size

How large are today's gangs? Again, the procedures for assessing gang size are highly variable. My police survey yielded size estimates, but of course they were dependent on whether one counted hard-core members, broader-core categories, fringe members, wannabes, and so on. Close to my own university, and therefore subject to the campus security's intelligence function, is a gang known as the Harpys.

They've been around for some years now, and their graffiti are abundant. Campus Security puts their numbers at 150—typical of a traditional black gang. The Los Angeles Police Department includes 800 names on the Harpy roster; five times the campus estimate. Another nearby gang, one of LA's largest, is listed in the LAPD as having 5,000 members in four adjoining police districts, with a complete roster of 15,000. For the same gang, the Community Youth Gang Services Program has offered estimates of from two thousand to six thousand.

I'm dubious about these LAPD figures, but important to the discussion is that in the absence of valid or at least consensual gang member definitions, such estimates are hard to assess. Recall from Chapter 3 that the majority of gang cities—and these would include many of the newer ones—reported relatively few gangs and relatively few gang members. Even in Los Angeles, the Sheriff Department's street gang manual lists 800 to 1,000 as the top size for black gangs. Chicago police figures yield an average of 900 per gang, with one "supergang" containing 9,000 (thus the typical gang would be far smaller than 900).[10]

Based on the available literature and the police survey, my guess would be that we have to entertain a series of different figures. Supergangs of the Chicago variety can indeed number into the multiple thousands of active members. Traditional gangs, as described in Chapter 3, typically contain from less than one hundred to several hundred active members (depending on their own cycles of gang activity, which vary greatly). Autonomous gangs, like the "spontaneous" gangs of the earlier era, probably range from ten to fifty members at a time, weighted toward the lower end. When these develop into linked branches, then numbers in the hundreds make sense, but we must be cautious about assuming that similarly named groups are in fact "branches." Many Crip gangs and Blood gangs that share those names are rivals, not friendly branches.

The gang proliferation of the 1980s was, as far as I can tell, composed more of autonomous than any other form of gang. If so, then the average gang size now is smaller than in the 1960s and therefore presumably more amenable to directed change. On the other hand, if poorly handled, there are lots of acorns out there that could become stable, traditional oaks.

Ages

There are only two logical directions for age ranges to go, up and down. Naturally, there have been claims for changes in both directions. However, both have not occurred. For many decades, the initial entry into gangs has been at around 11 years of age (initial, not typical), and so there is little room for change downward. Although some writers and officials decry the 8- and 10-year-old gang member, they haven't been in the business long enough to realize that we heard the same reports twenty and forty years ago. Sure, there are examples of the very young affiliating themselves with street gangs—remember the "unborns" mentioned earlier—but they are the exceptions, more important to media hyping then to criminological description or gang activity.

On the other hand, it is clear that the upper age range has expanded dramatically.[11] Both the crack and homicide studies among gangs in Los Angeles reveal the same change: The average offender in both cases was almost 20 years old, and the upper limits went into the forties. San Francisco, Honolulu, and Chicago all report similar patterns.[12]

Many of the newer gang cities, however, do not report gang members at the upper ranges. Adolescent and young adult members are the norm in these cities, old enough to be seriously affected by high unemployment rates for inner-city minorities, but not so old as to be contributors, as yet, to the persistent and pervasive urban underclass.

The older members seem more common in cities that have developed gang traditions over a decade or more. It's not that they join street gangs when in their twenties but that they now hang on to their affiliations longer.[13] This is certainly one function of unemployment rates among other variables, but not the only one. Issues of the urban underclass, discussed more fully in Chapter 6, have much to say about this "graying" of street gangs.

More important, perhaps, are the practical implications of these older, veteran gang members. They provide undesirable role models for new members and recruits. They are likely to be more sophisticated in crime, more likely to engage their companions in criminal pursuits including drug sales. Furthermore, these older members have more often gone through the tribulations and frustrations of inadequate education and job preparation, job loss and marital failure, and arrest and incarceration. Therefore, it will take more to bring them back into productive, prosocial roles than it will to divert their younger recruits into pursuits alternative to gang life. Where generational patterns of ganging have been allowed to develop, we have dug ourselves a deep hole.

Ethnicity

There still are some white gangs scattered here and there, for example, Armenians in two Los Angeles suburbs and small white gangs in Boise, Davenport (Iowa), Elgin (Illinois), Gadsden (Alabama), Green Bay (Wisconsin), and St. Louis. If I had not excluded bikers, stoners, skinheads, and their supremacist counterparts, the count would be considerably larger. Of almost eight hundred gang-involved cities, less than 10 percent have predominantly white gangs as broadly defined. Only five of these have a population of more than 100,000. The median size of white gang–predominant cities is 25,000, and most of these report three or fewer gangs. In other words, the white gang problem, although present, is not in any sense comparable to the size of the minority gang problem.

All the research reports and my own survey make it clear that the majority of American street gangs are black or Hispanic. In the city of Los Angeles, the LAPD's figures as of January 1990 show 45 percent Hispanic gangs (51 percent of all members), 41 percent black gangs (43 percent of all members), 9 percent Asian gangs (4 percent of all members), and 5 percent white (2 percent of all members). This yields an average gang size of 82 for Hispanics, 76 for blacks, 36 for Asians, and 18 for whites.

There have been changes in ethnicity accompanying the general proliferation of street gangs nationally. Those cities reporting the onset of gangs before 1971 listed Hispanics as constituting 40 percent or more of their gangs in 69 percent of the cities, whereas black gangs met this criterion (of 40 percent or more) in only 10 percent of the cities. But those cities acquiring gangs since 1984 report a Hispanic figure of 26 percent and a black figure of 46 percent. The old predominance of southwestern cities, including those in Southern California, yielded an early preponderance of cities with mostly Hispanic gangs. This has changed, so that cities with more black gangs are now the rule.

One thing that may have changed is interethnic violence. Although gang violence still is principally intraethnic—for instance, the LA data reveal 92 percent of gang homicides to be black on black or Hispanic on Hispanic—the proliferation of gangs may be increasing the numbers of cross-ethnic clashes. In perhaps the worst instance, thirteen people were killed in black-on-Hispanic battles in the district of Venice, California.

There may be a greater ethnic heterogeneity now than in the past, as well. Asian gangs are no longer Chinese or Japanese. Rather, they have expanded to homngeneous gangs of Koreans, Vietnamese, and Cambodians as well, with some reports of Mien, Hmong, and Eurasian groups.[14] Pacific Islanders—formerly only Filipino—now include Samoan and Tongan (as they long have in Hawaii). From the Western Hemisphere, one hears more of Haitian, Cuban, and Jamaican street gangs, whereas Guatemalan, Salvadoran, and Honduran groups have joined the array in the southwestern states, in some cases displacing the Mexican-American gangs in Los Angeles.

Ethnic homogeneity is still the common pattern within gangs. For example, Knox presents data from the Chicago Police Department showing average black gang homogeneity at 96 percent, average

Hispanic homogeneity at 94 percent, and average white homogeneity (nine cases) at 88 percent. However, reports were more common in my survey than I had expected of interethnic patterns, whereas in the past, ethnically pure patterns had been the rule, with very few exceptions. There are two changes here.

First, a number of cities—still a minority—report a few gangs that are ethnically mixed. Usually this is white plus Hispanic, but other variations were also mentioned. Meanwhile, a seemingly opposite trend was reported in some communities of interethnic gang rivalries. In the past, black, Hispanic, and Asian gangs fought only with other gangs of their own race. Now there are increasing reports of cross-race hostilities, fueled perhaps by ethnic residential successions and perhaps by drug territory problems. These changes are not prevalent, but they certainly merit some research attention not currently being provided. In an era of global acceleration of ethnic hostilities, there is no reason to expect gang members not to be affected.

Also not well studied have been the differences among the principal ethnic categories of gangs. Black, Hispanic, and Asian gangs can be differentiated because of their cultural and historical roots. In Chicago, Hispanic gang violence is far in excess of the nongang Hispanic rate, especially in the case of homicides. Almost half of Chicago's Hispanic homicide victims in 1989, for instance, were gang homicide victims specifically, mirroring a pattern over many years.[15] A similar predominance of Hispanic as compared with black rates has been reported in Los Angeles County for a number of years.

Irving Spergel compared Hispanic and black gangs in Chicago and related these to the effects of more recent immigration of Mexicans and Puerto Ricans into that city, in contrast with the far better established residence of Chicago's black population. As he sees it, Hispanic gangs have arisen in part to provide transitional organizational stabilities during the adjustment period, whereas black gangs may serve a stabilizing function as the black social structure is weakened by out-migration of middle-class blacks from ghettoized neighborhoods. In addition, Curry and Spergel demonstrated significant differences in Chicago among variables related to black and Hispanic gang involvement.[16]

Interestingly, this is the reverse of the ethnic successions in Los Angeles and other southwestern cities. In these areas, the Hispanic populations were in place long before the World War II and postwar immigration of large black populations. Spergel built an interesting conceptual analysis of the effects of segmented and integrated gang neighborhoods around his Chicago depiction. It would be interesting to see whether he would have to reverse it if immersed in the Los Angeles setting. My reading of his Hispanic picture sounds more like our black communities, and his black depiction more like our

Hispanic communities. In other words, ethnic differences may reflect migration patterns as much as they do cultural patterns.

As just one example of the need to make ethnic differentiations, Figure 4-7 illustrates the differences in gang–drug connections.[17] As suggested by press reports, the black connection is stronger.

The police survey, particularly as a function of the responses from Southern California, Texas, and other southwestern areas, strongly suggests the greater prevalence of the traditional, multigeneration gang in Hispanic communities. It also suggests, although by no means to the extent blared out by federal and state enforcement agencies, that black gangs have become more entrepreneurial and that Hispanic gangs remain more territorial. Crack, as the usual example, clearly is a more common commodity in the black community and has had more impact on the development of gang cliques or new, specialized gangs working the crack market. As I noted in an earlier chapter, many of my respondents were careful to specify that the black gang–crack connection was not the main pattern portrayed in the media.

It remains safe to say, as far as I can tell, that the black–Hispanic gang differences, though notable, still pale in the face of their structural and behavioral similarities. Differences are often more apparent than real because most scholars do not set out to compare ethnicities but, rather, to immerse themselves in one. There is, however, another ethnic pattern that, despite its own internal heterogeneity, seems to present a significant departure from black and Hispanic patterns. This is the Asian gang phenomenon.

In addressing Asian gangs, we are even more at the mercy of the media and of experiential data from police sources than is the case with black and Hispanic gangs. There are several reasons for the relative absence of professional research attention to Asian gangs. There have been far fewer of them, and their societal impact has been low. They have often taken root in nontraditional gang areas or areas where there is little history of gang research—Vancouver, British Columbia, Westminster and Garden Grove, California,[18] and Honolulu are examples of these.[19] Many Asian gang cities are among our newest and therefore not well noted. Finally, a point stressed by almost every police gang investigator assigned to Asian gangs: They are very hard to penetrate.

Asian gangs are exclusive and do not welcome intrusions from outside. Their Asian victims are more often uncomfortable with and distrusting of police and therefore fail to report crime to officials. Asian gangs are said to emphasize threats to the life and limb of their Asian victims. Finally, our country is almost devoid of Asian police, reporters, and researchers who could make useful contact with gang members.[20] But what we do know, or at least what we have agreed upon as known, presents a most interesting picture.

For instance, Pacific Islanders are numerous enough to be classed separately. A Los Angeles Sheriff's Department manual lists Filipino, Samoan, Tongan, Fijian, Guamanian, and Hawaiian gangs.

Chinese gangs differ in having greater organizational roots back home (mainland China, Taiwan, Hong Kong) in tongs and triads. New work by Ko-lin Chin on New York Chinese gangs and by Calvin Toy on San Francisco groups provides an inside look at these well-organized criminal groups, many of which stand midway between street gangs and organized crime syndicates.[21] Japanese and Korean gangs more closely resemble other American gangs. In all three cases, territoriality is said to be less residence based than prey based; that is, the gangs usually claim rights to the commercial areas that supply their criminal targets. Unlike blacks and Hispanics, many of the Asian groups are small, cohesive, and more materialistic and powerful, with less emphasis on intergang rivalries (although a few specific intergang incidents have been dramatically violent).

Pacific Islander gangs, about whom very little information is available, seem to have adopted more of the culture of black and Hispanic gangs, including graffiti, special dress and signs, and small territorial claims. This, at least, seems to be the pattern in Southern California. Salt Lake City, now reaping some negative side effects of Mormon missionary activity, is reported to be prey to both Samoans and Tongans, small but active in versatile crime patterns.

But as I noted, it's the Vietnamese that have captured press attention and seem most clearly to test their inclusion in the category of street gang. Very visible in Southern California, especially in towns like Westminster and Garden Grove, which resettled thousands of refugees following the American pullout from South Vietnam

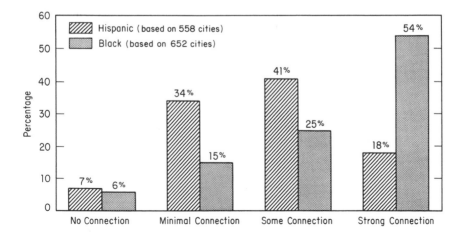

Figure 4-7. The Nature of the Black and Hispanic Gang–Drug Sales Connection.

(General Ky settled nearby), they also have reportedly been active in Atlanta, Houston, New Orleans, St. Petersburg, Washington, D.C., New York, Boston, Denver, St. Louis, Chicago, and Vancouver, B.C. Along with this dispersion goes a great deal of travel from city to city in the pattern so incorrectly attributed to the Bloods and Crips. The Vietnamese gangs are the most mobile of all, with each city providing a safe haven for travelers from the others.

These first- and now second-generation castoffs from their home country present the extreme form of characteristics sometimes attributed to all Asian gangs. They are highly mobile and relatively nonterritorial. They are secretive, loyal, and very difficult to penetrate. The age range is broad but concentrated at the lower ages. Most of them are male (with a very few, vicious copycat girls' groups reported).

Vietnamese gangs show less versatility in crime patterns than most, with an emphasis on a few property crimes. The offenses reported again and again are extortion of Vietnamese businesses and "home invasions" of Vietnamese families. These latter are robberies undertaken deliberately when the family is home (rather than burglaries of empty homes) because so many families distrust banks and hide their money at home. Home invasions often include violence or threats of violence to women and children in order to force them to reveal caches of money.

Violence among these gangs is uncommon, although a famous 1990 shoot-out at a gang funeral in a New Jersey cemetery signaled the severity of an exception. Violence, rather, tends to be instrumental, undertaken in pursuit of criminal gain: Threats, retaliations, warnings, and paybacks are not uncommon motives. But violence for its own sake, as part of intergang rivalry or as a component of spontaneous robberies, fights, rapes, and the like is less common than with most street gangs.

Thus, as can be seen, the Vietnamese gangs have developed a pattern that is not typical of street gangs, whereas that of most Asian gangs does less damage to my depiction of the American gang. I have no particular stake in including or excluding them but prefer to handle them as a test of the concept.

In regard to Asian gangs in general, it seems that two major generalizations are valid. First, they tend to be more deliberately criminal and to concentrate more on property crime than do black and Hispanic gang, or the occasional white gangs. Second, they are a very heterogeneous lot, so much so that further generalizations may be out of order. There are well over a dozen cultures from which these groups emanate, and they undergo socialization to American society and to American gang patterns in ways that reflect those cultures. When trying to understand and to intervene in these gang contexts, we would do well to acknowledge their heterogeneity.

Girls

These comments on Asian gangs could encompass the role of girls as well. There's very little literature on Asian girl gang members. If anything, the impression is that their role is even smaller than among other ethnic categories. Yet reports from Orange County, California, indicate a double exception: the existence of female Vietnamese gangs and their involvement in serious violence. The Garden Grove Police Department listed six autonomous girls' groups of ten to twenty members each, aged 13 to 20 years. They are described as aggressive and violent, mirroring some of the instrumental violence of their male gang counterparts.[22]

Now this may be another example of media hype, but it does raise the only issue I've come across that deals with today's girl gang involvement as compared with that of earlier decades (and described in Chapter 3). The little information available indicates that there has probably been little change in other respects.[23] The groups still exist, are more often auxiliary than autonomous, are smaller and fewer than male gangs, exhibit cafeteria-style delinquency, and are highly varied in their structures and crime involvement. They appear to be adolescent, not exhibiting the age expansion so notable among the male gangs.[24]

Certainly there is reason to expect more violence from girl gangs these days, to the extent that there has been an increase in violence among the male gangs, as outlined a little later in this chapter. In the most simplistic way, we could expect the violence of one to rub off on the other, but there may be a more sophisticated rationale, a gender difference in the handling of aggression generally, such that women attempt to constrain or only verbalize their anger, whereas men are given more leeway to act theirs out.[25] But in the specific context of the street gang, self-control of anger and aggression can be seen as a weakness. Gang girls, often victimized in their own families, join their group for both safety and acceptance. The trade-off is that they must exhibit their own acceptance of the group, specifically by acting out their aggression. This suggestion was clearly reflected in the study of an autonomous girls' gang in San Francisco.[26]

Thus aggression, which is socialized in normal settings as expressive behavior—crying, screaming, stamping of feet, and so on—becomes instrumental aggression in the gang, violence against rivals and victims and against fellow gang members when challenged. It follows logically that increased male gang violence over the years will be matched by proportional increases in female gang violence. Note, however, that this would be proportional, not equal. But let's keep even this in the proper context. Gang girls are still in the minority, so their violence, though obviously troublesome, does not attain the social problem salience of male gang violence. Los Angeles Sheriff's

Department files show just 2 percent of the suspects in 1,346 homicides since 1979 to be females.

New Gangs? Violence, Drugs, and Entrepreneurs

Granted that there are enormous numbers of gang cities as compared with earlier times, but as previously noted, many of them are facing relatively small gang problems. How much more serious is the situation now? With respect to criminal involvement, I want to consider three issues, quite related in the public mind and at the heart of the media/enforcement hyping of the American street gang. These three are changes in violence, the purported equation of gangs with drug distribution, and the emergence of a new form of gang, the entrepreneurial organization. I'll treat the three separately, but clearly they are related issues.

Modern Gang Violence

Within this category let's consider three subissues: the special nature of gang violence, its prevalence, and its increase. We have the best information on the first of these.

Gang Differences. Back in the mid-1960s, much was theorized but little was known about the differences between gang and nongang delinquents. The first important breakthrough was the analysis in the 1960s in Chicago by Short and Strodtbeck. What was needed to replicate the gang–nongang differences found then was to select gang and nongang members from the very same areas of residence and to assess the gang-versus-nongang offenses in numbers, types, and eventual dispositions.

Thus I was very pleased to see Paul Tracy's analysis in 1979 of the gang data in Philadelphia.[27] It concluded, as one might expect, that there were major differences in official arrest rates between gang members and other arrestees. Among juveniles, the ratio was almost three gang member arrests to one nongang arrest. At the adult (18 and over) level, the difference was even greater, a ratio of more than five to one. When self-reported offenses were used instead of arrests, the ratios were very similar (with, of course, much higher numbers).

This sort of analysis is unusual. Most gang research projects do not look at nongang offenders at the same time. A few other reports have appeared and typically find significant differences in the direction of greater gang effects. This includes the Denver and Rochester longitudinal studies cited in an earlier chapter. In Los Angeles, in the context of a project aimed at assessing the validity of police-reported gang violence rates, we found some major gang-versus-nongang differences in the character of the homicide incidents, using data from 1978 to 1982. Table 4-1 shows the pattern for setting and participant variables in the LAPD data, and Table 4-2 shows the patterns in the LASD data.[28]

The gang and nongang homicide cases were drawn from the same LAPD and LASD areas; thus they were geographically controlled. The only other control was the limitation to suspects between the ages of 10 and 30, roughly the effective limits of gang membership. For the statistically sophisticated reader, I can report that a discriminant analysis yielded quite high classification success, cross-validated by split half procedures and confirmed with logistic regression and quadratic classification analysis. For others, it is sufficient to note that each of the differences between gang and nongang figures is beyond chance, that overall the gang homicide is quite a different affair from the nongang homicide.

Was this an accident of time, a peculiarity of the then-peak period of LA gang violence of 1980? We repeated the analysis using 1988 and 1989 cases from the same five police station areas. The patterns were very similar to those from the earlier period and to another set of data analyzed from 1984 and 1985. The differences are stable, overcoming any deficiencies inherent in the police-recording processes. Most important here is the conclusion that it is not just the amount of gang-member crime that differs but that its character differs from that of nongang crime. This is at least true with respect to homicide and also with respect to the nonfatal violent incidents we sampled from the Sheriff's Department and from five other cities in California, north and south. Those offenses—assaults, rapes, shooting into occupied homes, robberies, arson—again exhibited the patterns found in Tables 4-1 and 4-2. These stable differences, over a variety of violent offense categories and over a variety of locations, indicate, I fear, that we have come a long, not-very-nice way from the more benign gangs that I and others studied some decades ago.

Levels of Violence. If gang delinquency and other crimes are qualitatively different from each other, then the proliferation issue assumes greater importance because the differences may be widespread. But here it is wise to caution the reader again about the great variety among gangs and among cities. Most cities, fortunately, are not like Los Angeles. In Honolulu, for instance, it is reported that only 27 percent of juvenile gang and 25 percent of adult gang cases involve serious (FBI, Part I) offenses. Indeed, 30 percent of all juvenile cases are status offenses (chargeable only against minors: runaway, truancy, possession of alcohol, etc.). These gang members sound like those of old.

Chicago analyses yield several cautions. Only 10 percent of all homicides in the mid-1980s were gang related (but remember that Chicago uses the narrow, motive-related definition). Lawrence Bobrowski, the superb statistical analyst in the Chicago Police Department, reveals for 1986 through July 1988 that serious (FBI,

Part I) gang crimes in that city comprised about 1 percent of the city-wide total. Within that, homicides got as high as 18 percent of the total, although the average was a far lower 8 percent, and serious assaults averaged 4 percent of the city total.[29] There is, in other words, a lot of action that is unrelated to gangs, despite the impression one gets from the media. Finally, more recent research reveals that just four of Chicago's most notorious gangs accounted for 55 percent of the gang homicides and for 69 percent of all recorded gang incidents in the period from 1987 to 1990, with gang violence concentrated in only a few city areas.[30]

In Los Angeles County, gang homicides now account for between 35 and 40 percent of all homicides countywide. In fact, nongang homicides have been going down for the past few years, so the gang contribution is quite an anomaly. Yet despite this, in the city of Los Angeles in 1989, gang homicides (303 of them) constituted only 4 percent of the 7,332 serious offenses attributed to gangs by the LAPD (and no count is kept of all the "nonserious" offenses—thefts, vandalism, drug sales, and so on). Countywide, of course, the figures are far greater. Interestingly, despite the relatively similar numbers of black

Table 4-1. In the Jurisdiction of the Los Angeles Police Department

Setting	Gang (N = 135)	Nongang (N = 148)
Gang homicides more often occur in the street	49%	34%
Gang homicides more often involve autos	64%	49%
Gang homicides more often involve guns	83%	68%
Gang homicides less often involve knives	24%	37%
Gang homicides more often include unidentified as sailants	23%	10%
Gang homicides more often involve fear of retaliation	33%	9%
Gang homicides more often involve injuries to other persons	23%	14%
Participants		
Gang homicides involve a higher average number of participants	6.96	3.77
Gang homicides more often involve victims with no prior contact with their assailant	49%	27%
Gang homicides are more likely to involve clearly gang victims	40%	2%
Gang homicide suspects on the average are younger	19.40 yrs.	23.68 yrs.
Victims on the average are younger	23.67 yrs.	31.06 yrs.
Gang homicide suspects are more likely to be male	94%	84%
Victims are more likely to be male	95%	89%
Gang homicide victims are more often Hispanic	53%	39%

and Hispanic gangs, Hispanics now are victimized at almost twice the rate. This belies the usual media portrayals of drug-driven black gang homicides having a corner on the market.

Los Angeles and Chicago, in that order, are the most gang-ridden cities in the nation. If we look away from them to obtain a national picture, I can report several pivotal items about prevalence:

1. The number of cities now reporting gang problems—still an underestimate because we haven't surveyed all cities—is almost 800. The figure derives from the 261 phone interviews, the random sample of 60 cities with populations between 10,000 and 100,000, and our survey of over 1,100 cities concerning their experience with gang migration.

Table 4-2. In the Jurisdiction of the Los Angeles Sheriff's Department

Setting	Gang (N = 226)	Nongang (N = 220)
Gang homicides more often occur in the street	48%	14%
Less often at residence	24%	53%
Gang homicides more often involve autos	66%	56%
Gang homicides more often involve guns	80%	60%
More often involve other weapons	31%	22%
Involve a higher average number of weapons	2.23	1.68
Gang homicides more often involve additional offenses	72%	52%
Gang homicides with additional offenses are more violent, such as other homicide charges	45%	22%
Assault with a deadly weapon	57%	39%
Less often include robbery	20%	34%
Gang homicides more often involve injuries to other persons	30%	10%
Gang homicides more often include unidentified assailants	19%	7%
Gang homicides more often involve fear of retaliation	33%	10%
Participants		
Gang homicides involve a higher average number of participants	8.96	3.59
Gang homicides more often involve victims with no prior contact with their assailant	54%	24%
Gang homicides are more likely to include clearly gang victims	47%	4%
Gang homicide suspects on the average are younger	19.16 yrs.	24.02 yrs.
Victims on the average are younger	23.5	29.0
Gang homicide suspects are more likely to be male	97%	87%
Victims are more likely to be male	92%	82%
Gang homicide suspects are more often Hispanics	74%	30%
Victims are more often Hispanics	83%	39%

2. Among the 800 gang cities, 52 represent a serious gang violence picture with ten or more homicides per year (1991). These
cities report a total of 2,600 gangs and 200,000 gang members. That's
a lot, by any measure, no matter how varied their definitions of gangs
are.

3. In each of the 800 gang-involved cities located by the migration survey (including my original 261), we asked about gang-related
homicides for the year 1991. We obtained a total number of 453 non-
homicide cities. Another 247 had fewer than 10 homicides, but when
adding them to those having 10 or more (the cities used in item 2
above), we got a rough total of 2,166 homicides attributed to street
gangs in 1991 in our surveyed cities. These figures omit the cities that
did not report their homicide figures to us.

So many gangs and so many gang members contributing so many
homicides to our already violent nation must be counted as a major
social problem. Recall that homicides probably represent less than 5
percent of serious gang offenses (using Los Angeles as the base for
this) and certainly less than 1 percent of all gang offenses. Thus total
gang offenses are a very, very large number. It would seem clear,
would it not, that the time for enforcement and media posturing had
passed and that the time for seriously considering something other
than knee-jerk responses has come.

Increases in Violence. Beyond the unsystematic collection of anecdotal reports, there is little solid evidence to support the assertion of
an increase in gang violence over the last decade or two. Our survey
report of homicides in 1991 speaks to levels, not increases. It's all
inferential beyond that.

However, a few specific locations have provided relevant data. In
Chicago, an increase emerges across the up-and-down cycles. There
was one peak in gang homicides in 1981 and then another recent rise
to 98 in 1990 and a new high of 125 gang homicides in 1991 (using
the narrow definition because the broader one might take the figure
closer to 200). At least some of the increase has been attributed to the
greater lethality of weapons now in the hands of the offending
gangs.[31]

The most solid evidence of violence increases comes, of course,
from California and more specifically from Southern California. The
state attorney general's office, through its Bureau of Criminal
Statistics and Special Services, compiles relevant criminal justice data
from cities and counties throughout the state. Between 1980 and
1989, these figures included "known contributing circumstances" to
homicides. Whereas drug-related events, those given so much publicity, increased from 5 to 10 percent over this ten-year period, gang-
related events rose from 5.5 percent to 18.5 percent. Only arguments
(domestic and otherwise) exceeded the gang figure. Black and
Hispanic gang contributors were about equal, as compared with the

drug-related events, for which black figures were twice those of Hispanics. The crack influence is on drug-motive, not gang-motive murders.[32]

In Orange County, an affluent area with a seldom-discussed backwater of poverty pockets and traditional Hispanic gangs, recent increases in Vietnamese gang violence and the growth of Hispanic gangs have brought disheartening awareness to the area. In one year alone, between 1990 and 1991, the district attorney's office saw gang-member charges for homicide rise from 26 to 43. The county now records 200 gangs, mostly Hispanic, and 15,000 gang members in a constantly accelerating pattern.

In East Los Angeles, there has been a structural change to a more institutionalized and entrenched form of Hispanic gang with increased deviance in general, including concurrent but not causally related rises in violence and drug involvement. Part of the increase in violence is attributed to better weaponry and part to the challenge from newer cliques to outdo their predecessors.[33]

One manifestation of this has been the increase in "drive-by" shootings, although drive-bys are certainly not new. In fact, they were quite common among our four observed black gangs in the early 1960s, and they also were common in five Hispanic gangs when we observed them in the mid-1960s. One of our research assistants was present at a Slauson drive-by in Gladiator territory—five shots fired into a large gathering to which the assistant had been sent to check out the "rumor." Another, a drive-by shooting, is what drove a badly shaken gang worker into our clutches to initiate work with the Latins.

Because drive-by shootings have become a media hallmark in the more recent period and they admittedly are more frequent than they used to be, an interesting descriptive analysis was undertaken by two physicians in the Department of Emergency Medicine at Los Angeles County Hospital. This is where many victims end up if they are still alive when the ambulance comes to the scene.[34] Using countywide data, the analysis shows a rather steady increase in gang-related homicides per 100,000 persons (thus controlling for population increases) but a decline in nongang homicides. The gangs are bucking the overall trend. Figure 4-8, adapted from this paper, shows this pattern clearly, in which the gang changes mirrored overall homicides up to 1982 and thereafter failed to do so.

Within that context, the authors then reviewed drive-by statistics in the city of Los Angeles for 1989 through 1991, limited to these years because the LAPD started compiling the data only in 1989. Table 4-3, again taken from that paper, provides the data. Note the steady increase in almost all the indices provided. Table 4-3 does not refer to gang events only, but it is heavily affected by them nonetheless. The real point has to do with a pattern that is increasingly prevalent, visible, newsworthy, and difficult to control. The authors point out that

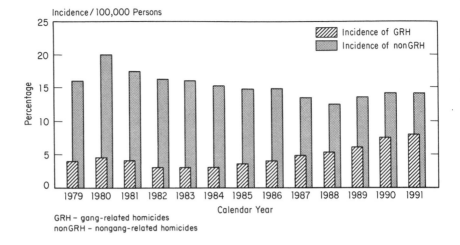

Figure 4-8. Los Angeles County Homicides 1979–1991.

in 1991, countywide (the LASD started collecting data in 1991) there were eleven drive-bys per day. The authors were particularly careful to note thirty-two "pediatric homicides by drive-bys" in 1991 in the city and four retaliatory gang shootings in local hospitals.

Finally, I come to the data I offer with the least pleasure. I have been asked to give lectures on gangs in many states and overseas as well. I enjoy these opportunities until it comes time to display Figure 4-9. I've lived in Los Angeles for over thirty years now and am very fond of it. Its topography ("ski and surf"), climate, social openness, and fascinating ethnic heterogeneity all make for a good lifestyle. Its cultural offerings—the symphony, theaters, museums, and ethnic anomalies like Olvera Street, Venice Beach, and West Hollywood—match or exceed those of most cities. I've lived here in solid black, racially mixed, and predominantly white residential enclaves. They all were satisfying. So I take pride in Los Angeles, and I like to talk about it. But being a gang researcher here certainly tarnishes the halo. Figure 4-9 tells its own story.

Table 4-3. Drive by Shooting (DBS) Incidents and Deaths, City of Los Angeles, 1989—1991

Year	No. of DBS/Year	Ave. No. of DBS/Day	No. of Victims of DBS	No. of Deaths by DBS	Percent. of Deaths by DBS
1989	1,112	3.0	1,675	78	7.0
1990	1,263	3.5	1,824	122	9.7
1991	1,543	4.2	2,222	141	9.1

The data come from the Sheriff's Department, which gathers them annually from all enforcement jurisdictions (numbering more than fifty) in a format little changed over the years. The pattern was foreshadowed in the earlier tables, but the absolute numbers of homicides stand out (remembering, as always, that the broader, member-defined definition is used in Los Angeles). Consider just 1992, in which there were 803 deaths, 803 families in mourning, more than that number facing murder trials for their offspring, and more than that number facing the futility of prison life and its aftermath.[35]

This is, in short, a terrible picture, a terrible situation, and a source of considerable anguish for a dedicated Angeleno like me. I just don't like this part of the lectures. In the past decade, the usual up-and-down cycle of gang activity simply has not taken place, though past experience says a downturn is long overdue. Close to 40 percent of all homicides in Los Angeles County are now gang related, and still the county has not gotten its act together to work on this problem except in the most superficial, after-the-fact fashion. I shall, of necessity, return to this issue in Chapter 5.

Now, quite obviously, the Los Angeles area is not the nation. In the gang arena, it is an excrescence that should warn all other cities to get to work on gang prevention. Keep in mind that Chicago is not all that far behind and that Washington, D.C., is the nation's murder capital even without a notable street gang problem. The Los Angeles area probably contributes in the neighborhood of one-fourth of the country's gang homicides and perhaps more of its gangs. But the lesson of the increase in so many U.S. cities, despite the few social programs and intensive suppression programs already under way, needs to be considered. My survey revealed very few towns and cities experiencing a decrease in gang problems; the trend line is definitely an accelerating one. The current fashion of blaming it all on drugs isn't helpful, and in most areas it's probably just plain wrong. It misleads our efforts at intervention. Street gangs, for their members' sake and that of their communities, deserve our attention as a problem in their own right.

Crack and "Drug Gangs"

As I view it, these last two sections of Chapter 4 are quite special. I'm going to repeat myself, add a bit of new material, attack some of my enforcement and academic colleagues, and generally make a pest of myself. But in the end, I hope to provide some clarity. I've made the point in earlier pages that with some notable exceptions, street gangs are not in control of drug distribution and, in particular, not in control of crack distribution, as has so often been alleged. In fact, I've over-stressed the point because there has been so much public attention to the gang–crack connection.

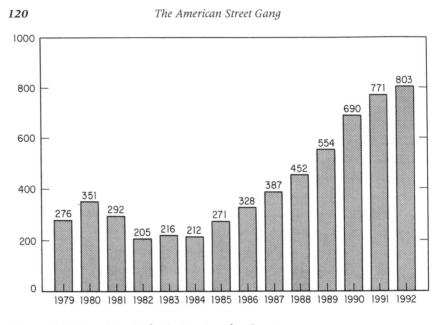

Figure 4-9. Gang Homicides in Los Angeles County.

In all fairness, I must warn the reader that in my view—and it is a strongly held view—there is an unfortunate connection between the official and the public acceptance of research and the "flashiness" in the way that this research is presented. The belief in a strong gang and drug sales connection is fueled, in part, by academic writers who play to the dramatic aspects of the purported connections. The more journalistic these writers become (and the less careful they are about handling and presenting their data and data collection procedures), the more they capture their audience's attention at the expense of accuracy. The drama seemingly inherent in street gangs and drug trafficking is seductive and, like all drama, leads by emotions more than intellect.

Claiming the Connection

Let's look at some additional examples of the claims to street gang involvement in drug distribution, both from public agencies and from my academic sources.

- Former Chief Daryl Gates, quoted in a December 14, 1990, *Los Angeles Times* interview, commented,

   Well, I think that the drugs are there and probably the gang violence would not occur if they weren't all on drugs. They are already, as the Sheriff says, driven by a sociopathic mentality that kind of permeates the group, and they have a few beers and snort some coke and they are on their way. "Why don't we have some fun? Let's go shoot somebody."[36]

- The Sheriff's Department captain and gang expert, the one with the "hoodlums" definition of gangs, told his trainees, "Gangs and drugs are almost the same problems."[37]
- The chairperson of the 1991 New Mexico Governor's Organized Crime Prevention Commission reported, "We are no longer talking about juvenile delinquents when we speak of the street gangs of today. We are talking about well-organized, drug-dealing, profit-motivated and dangerously armed young hoodlums who commit vicious crimes."[38]
- A report to the president from the U.S. attorneys and the attorney general of the United States concludes:

> California is home to one of the most dangerous and menacing developments in drug trafficking, the large scale organized street gang.... The Los Angeles gangs are radiating out from the areas where they originated—up the West Coast as far as Seattle and Vancouver, into the heartland as far as Denver, Kansas City, and Chicago, and even to cities on the East Coast.... One of the most frightening aspects of California street gangs is their willingness to direct their violence at each other, at the police, at members of the public—at anyone who stands in the way of their operations.[39]

- Information from the U.S. General Accounting Office in 1989 reported that "some law enforcement agencies have estimated that the Bloods and the Crips, two gangs based in Los Angeles, control approximately 30 percent of the national market in crack cocaine." One has to wonder which Blood and which Crip gangs these are, and how on earth one can establish that they have a 30 percent share in a market that is inherently unmeasurable.[40]

Unfortunately, this theme too often is picked up and repeated rather uncritically by some of my colleagues. In an article unusually dependent on *Time, Newsweek,* and the like, the chair of a criminal justice department provides the following florid statement:

> The unspeakable horror of the deadly drug plague raging in America's cities has transformed them into combat zones. In more than a dozen major metropolitan areas, rival factions of automatic weapon-toting, drug dealing, young black gang members wage gruesome guerrilla war for control of the thriving drug trade. As business booms, the body count rises.[41]

Even an astute and productive New York drug research group can't resist passing along the official line about Los Angeles in the direct face of our controverting data:

> While free-lance relationships among dealers and sellers are common in Los Angeles, two loosely organized black gangs, the CRIPS [*sic*] and the Bloods, have become heavily involved in selling heroin, cocaine, and

crack..., although Klein, Maxson, and Cunningham...found only a loose
overlap among gang members and drug sellers. The Drug Enforcement
Administration...believes that these two Los Angeles gangs control and
distribute crack throughout most West Coast cities and even in the
South.[42]

I want now to give special attention to the work reported by
Jerome Skolnick because Skolnick's conclusions about the gang–crack
connection and the emergence of "entrepreneurial gangs" in control
of crack distribution were widely and uncritically adopted as revealed
truth by many enforcement officials, including major players in the
National Institute of Justice. I want to be clear that his hypothesis
regarding new entrepreneurial gangs is an attractive one, about
which I have already supplied some empirical support, that is, the
existence of "drug gangs." But the equation of these groups with
street gangs and the suggestion of their national franchising operations on the basis of interviews with a few dozen convicted felons are
not a good basis for establishing general facts or public policies.

In the next pages, then, I offer my own appraisal of this
situation.[43] The reader may recall that in our Los Angeles research we
found little evidence of sophisticated gang organization in these largely street-level sales incidents and concluded that crack distribution
seemed to remain largely, though by no means exclusively, in the
hands of the regular narcotics distributors.

In contrast, Skolnick's 1988 interviews with thirty-nine inmates
at four California correctional institutions and with forty-two police,
narcotics, and correctional officials revealed different patterns. He
found that gangs dominated the drug trade in Northern California. He
termed such gangs *instrumental* or *entrepreneurial* and found them to
be highly structured organizations, strictly "business" operations
whose primary purpose was distributing drugs. He distinguished these
from the "cultural" gangs found in Southern California, exemplified
by traditional, or neighborhood-based, identities, especially as
expressed by the Chicano gangs in Los Angeles. These are not organized for the specific purpose of distributing drugs. Black gangs in
Southern California are reported to be loosely modeled on Chicano
gangs but are less rooted in a cultural base and are moving toward the
instrumental model "as the values associated with drug marketing
come to dominate members."[44]

Sources of information for the two studies differed. The Los
Angeles data came from a slightly earlier period, from arrest records
and police gang identification rosters that may have underrepresented
gang involvement. On the other hand, the impressions we gathered
from informal interviews with both gang and narcotics officers during
our study are not inconsistent with Skolnick's reports. They simply
were not supported by the data we gathered.

Moreover, in-depth interviews with gang members in other studies directly refute Skolnick's contentions. Both Waldorf's group in the Bay Area and Quicker's interviews with 61 current gang members and/or drug dealers in South Central Los Angeles and contiguous areas found no support for Skolnick's descriptions. "We found no evidence of the existence of, or a transition to, entrepreneurial street gangs taking place."[45] "In general, it is our impression that most drug sales by gang members is [sic] not particularly well organized and nothing like Skolnick's descriptions, but there are some groups that are highly organized" (personal communication from Dan Waldorf, 1991).

Skolnick's inmates may not be reliable informants on street drug dealing, or his sample may be highly selective. He does not seem to advance much skepticism about their reports and finds independent validation only from his law enforcement sources. Based on the Waldorf and Quicker studies, as well as those from other researchers such as Fagan, Huff, and Hagedorn, I continue to be skeptical of Skolnick's gang typology and related findings about the structure of street gang drug sales. I make this point because Skolnick recently released a report through the California Department of Justice on gang organization and migration, utilizing the same methodology. He states that his group continued their study in the summer of 1989 but concentrated on gang migration issues. He does not report on the nature of his interview sample (but refers consistently to "inmates"), number of subjects, their distribution along ethnic, geographic, or age dimensions, or whether these were reinterviews of the 1988 sample of thirty-nine inmates versus new subjects.

Skolnick's report is the most current, relevant, and extensive piece yet available on gang migration, so I review his findings in detail, yet cautiously. Note that he deals exclusively with the drug sales aspect of gang migration and does not speak to migration of any other type. Moreover, it's unclear whether these reports are of residential relocation or temporary visits.

Scope of Migration

Skolnick states that the information provided by his respondents confirms reports originating from the FBI (but cited in *Newsweek* magazine) of Blood and Crip operations in forty-six cities. "It is difficult to overstate the penetration of Blood and Crip members into other states."[46] His accompanying footnotes list thirty-two cities, sixteen of which are in California, and seven other states that have no city locations specified. In addition to wondering about his respondents' depth of familiarity, this lack of specificity makes it difficult to cross-check Skolnick's lists against others now available. Moreover, he does not question whether these might be "copycat" Crips and Bloods; he assumes them to be Los Angeles–based migrations and therefore may well be overstating the case. He does not address how many members or how many gangs might be involved; his data do not speak to the magnitude of gang migrant–initiated drug sales in destination cities relative to local contexts of drug or gang activity.

In San Francisco, the Waldorf team's contact with eighty-four gangs yielded only three with any Los Angeles gang connections, and these were "very tenuous."[47]

All the Los Angeles–area Blood and Crip respondents in Skolnick's study (number and source of documentation of gang membership not provided) reported that "traveling" takes place in their sets, and Skolnick states that migration has become an institutionalized aspect of drug dealing. But no basis for estimates of the proportion of LA gang members involved is offered.

Which Gangs Migrate?

Skolnick finds that only Southern California gangs migrate; Northern California gangs do not. He posits a "cultural resources theory" to account for why the "horizontally organized" cultural gangs do travel and the "vertically organized," entrepreneurial Northern California gangs do not. Note that the attribution of vertical-versus-horizontal organization contradicts much previous research on gangs and much of the new research as well.[48] Skolnick suggests that there are various drug-dealing benefits of cultural gang membership that have significantly increased over time. Yet it seems that only the black gangs in Los Angeles, which he has said are less culturally influenced, can capitalize on these benefits by migrating to establish new drug markets. They seem to have accomplished this feat by transforming their relationships to drugs into instrumental capacities.

Although Skolnick's attempt to advance theory in this area is commendable, it is difficult to break down his exposition into testable hypotheses with operational indicators. In his report, cultural resources are variously increasing and decreasing, to both the benefit and the detriment of migratory activity related to drug sales. Instrumental resources seem both present and absent.

Motivations to Migrate. Gang members are reported by Skolnick to migrate in pursuit of higher profits (the LA area is stated to be saturated) and to exploit their perceived competitive advantages in other drug markets. However, the decision to migrate does not result from strategic planning but is activated from ordinary family visits. Over time, the operation expands, becoming more sophisticated "as Bloods and Crips begin to develop an extensive network of contacts in cities and counties throughout the nation.... An elaborate and well conceived transportation plan must be developed...drugs must be dispatched to the new market and proceeds returned to Los Angeles."[49]

Rebutting the Connection

Let's turn to some other researchers, who seem more in line with our Los Angeles conclusions than with Skolnick's. Contrary examples, and indeed there are some, will be noted in the section on drug gangs.

In Chicago, the Blocks undertook an intensive analysis of homicides and concluded that

> only eight of the 285 street gang-motivated homicides between 1987 and 1990 were drug related. Five of these, all of which occurred in 1989 or 1990, were related to the business of drugs. As Klein, Maxson, and Cunningham...found in Los Angeles, the connection between street gangs, drugs, and homicide is a rare one, and not a major explanation of the rapid increase in homicide in the late 1980s.[50]

The broader review by Collins, citing both Moore's and Fagan's work, makes the point nicely about what effective drug distribution requires:

> The above does not argue that drug trafficking lacks formal organization, however. The drug distribution system is quite complex and often involves formal organization. The system is not monolithic—either vertically throughout the distribution levels or geographically. Multiple organizations participate at wholesale and retail levels and in different areas. There is a tendency to ignore this multiplicity and to see fully organized conspiracies where none exist. Exaggerated rhetoric about gang control of drug trafficking is an example.[51]

Closer to home, Moore's comment helps as well, because no one would dispute that gang members sell drugs—some more, some less, some not at all. She says with respect to her Mexican-American groups, "The gang itself is not acting as a unit to deal drugs, but individual members of the gang are dealing drugs, and drawing on one another as partners, completely outside the context of the gang as a whole."[52] In corresponding black gang sections of Los Angeles, Quicker and his ex-gang coauthors and informants draw basically the same conclusion, that crack selling is not a gang enterprise: "Street dealers represent the bottom rung of the drug organizational hierarchy. While many of these individuals are street gang members, a significant proportion are not."[53]

Two more items on the connection with homicide: First, we replicated our earlier data collection from the homicide investigations in the five most crack- and gang-involved sections of Los Angeles County. Using the same forms of data and the same analysis procedures for the years 1988 and 1989, we obtained substantially the same results. Yes, there is drug involvement in a number of the homicides, and drug motives are found in about one-third of the cases. But these connections are even more true of nongang than gang homicides. It is this comparative finding that is critical. There's no denying some gang–drug connections in homicides, only the special connection, above and beyond what is driven by drugs generally.

Second, the Centers for Disease Control, at the explicit request of the Los Angeles Police Department, undertook similar analyses using 2,162 homicides between 1986 and 1988. They corroborated our findings, with even less support for the purported connection. Five percent of gang-motivated homicides involved narcotics, as compared with 18 percent of other homicides. Regardless of age and regardless of location, the results were consistent: "The results of our investigation...do not support the popular theory that a substantial portion of homicides are attributable to gang involvement in narcotics trafficking...gang-motivated homicides do not tend to involve narcotics trafficking (or victim drug use), and narcotics-motivated homicides do not tend to involve gangs."[54] This report, as noted, was based on LAPD data. Similar data from the Sheriff's Department in 1992 showed that in cases in which motives could be discerned, 85 percent pertained to gang rivalries and only 3 percent were related to narcotics. A ten-year analysis of Boston data reported by Walter Miller in a personal communication revealed that less than 10 percent of all violent gang incidents, including homicides, were drug related.

The Structural Issues

Several of my earlier comments speak again to the unsuitability of gang structure and organization for effective drug distribution (not just sales). Right in Skolnick's backyard in San Francisco, Waldorf concludes that

> most gangs do not have the skills or knowledge to move to other communities and establish new markets for drug sales.... Lest we forget who are the youth who make up gangs—they are nearly always young people who have few resources, who leave school early, have limited skills and are often just as unsuccessful in their illegal ventures as they are in legal ventures.[55]

New York City, our largest, is also a city with enormous drug use and distribution activity. New York is also one of the original, traditional sites of widespread street gang activity. What can we learn from New York about our drug and gang connection issue?

Despite the enormous output of various drug researchers there, the lesson comes from omission. With one exception, a brief mention in an article by Paul Goldstein, descriptions of the organization of drug distribution in New York are utterly devoid of any mention of street gangs. Whether the drug is heroin, powder cocaine, or crack, the absence of gang observations is striking. Jeffrey Fagan, a major contributor to both the drug and gang literature, has no difficulty describing the causal independence of the two problems in three cities, including New York.[56] Yet in his description of New York alone, the "g word" is missing. Organizations, *ad hoc* groups, decentralized distribution systems, and incipient organized crime groups all are phrases brought to bear in Fagan's recent review of New York drug selling, but *gang* is absent, clearly quite deliberately so.[57] Drugs, including crack, are a constant, pervasive feature of many borough areas, but street gangs are not needed for the business. And New York does have gangs, fewer than in the old days, it is said, but they are in no way prominent in the business.

Why? One reason is that the gangs, as described by the New York Police Department, are small (an average of twenty-five members), loosely knit, unconnected local groups—much like the autonomous gangs appearing in many of the country's newer gang cities. Gang members are involved in drug sales, but not gangs as units. The department states:

> In New York City, drugs are controlled by organized crime groups. Young, weak, undermanned and poorly organized street gangs cannot compete with the older, more powerful and violent groups. The fragmented street gangs do not have the network or the power to distribute or control drugs on a large scale.... The primary difference between a drug organization and a youth gang is that in a drug organization all members are employees while youth gang membership only requires affiliation. We do not see our youth gangs becoming drug organizations.[58]

So let me repeat it. Drug distribution requires good organization, and most street gangs are not well organized.

Drug distribution needs hierarchical, dependable leadership, whereas gang leadership tends to be limited by age, ephemeral, and functionally narrow. Leader reputation does not necessarily translate into leader influence.

Drug distribution requires interpersonal loyalty; street gang loyalties can be very strong in feeling but often are manipulated by good police work.

Drug distribution requires focusing on the business and not becoming involved in irrelevant criminal activities; street gang crime features "cafeteria-style," opportunistic endeavors along with the more appropriate planned events.

Drug distribution requires a cohesive organization, with the membership carefully controlled, whereas street gangs normally do not maintain the level of cohesion needed for a sustained market operation.

But members can easily become involved. Cliques can form around drug sales. New groups can form for the purpose of selling drugs. Outside distributors can recruit gang members and nonmembers in a franchising operation, but seldom can they recruit an existing street gang as a unit and convert it to a branch operation.

We need to keep in mind the considerable variety of forms that 1980s and 1990s street gangs can take. Like amoeba, this form can flow in a number of directions, toward higher and lower cohesiveness, broader and narrower focus in crime, more and less defined leadership, greater and fewer connections with crime figures, and so on. Thus some street gangs are hard to distinguish from drug gangs, and some gang alliances seem or become similar to criminal or drug distribution networks. The emphasis is on *some*. Most street gangs can't make it, no matter how convenient that would be for enforcement officials and the media.

There are good examples of gangs dealing drugs that sometimes drive the popular image. Padilla's description of a Puerto Rican group in Chicago is an example, and Williams's depiction of the Dominican "Cocaine Kids" is another. Taylor's depiction of Detroit dealers can be added, although I have reservations about it that I will get to later. Hagedorn's picture of selected gang founders in Milwaukee presents elements of the pattern. Some Jamaican posses offer an outsized example, as in Kansas City.[59] Certainly there are a number of Los Angeles gangs more heavily involved than are most of their counterparts.[60] These groups buck the trend, and we certainly need research to help us understand why and what this portends for the future.

We thus need to remain alert to the easy overstatements about street gangs as "natural" mechanisms for controlling drug sales. Scott Decker used interviews with scores of St. Louis gang members specifically to test the hypotheses of Skolnick and others, on the one hand, and mine, on the other hand (with references as well to Fagan and Hagedorn, among others). Decker reports a set of results fully in accord with the position I've taken here:

- "The responses with the largest frequencies...do not reflect instrumental benefits of gang membership."
- "Only 15 [of 99] gang members we spoke with told us that this [an opportunity to sell drugs] was an important reason to join their gang.... For most gang members, drug sales were simply part of a larger overall group of activities."
- "Most gang members indicated that their gang did not require members to sell a certain amount of drugs."
- "...drug sales by gang members are common, but...the gang does not play a central role in such sales."
- "We found no evidence that new members were recruited because of their entrepreneurial skills."

- "Few such commitments to instrumental concerns were to be found. Indeed, as others have found...the gang is unsuited for the tasks of building consensus among its members and organizing its activities."[61]

Variations on the Drug Gang Context

Street gangs are not a single, monolithic structure. But I haven't said much about the so-called drug gangs because so much less is known about them—ethnographic research is only now reaching publication.

The multigang contexts in which drug gangs have emerged are quite varied and quite interesting. A sampling of comments from police respondents in our one hundred largest cities illustrates this variety. These are reports from a particular time slice—late 1991 or early 1992—so that there may well have been changes by now.

- Albuquerque: traditional street gangs, drug gangs, and some combinations of the two.
- Baltimore: organized drug distribution groups, plus Jamaican "posses," but no street gangs (although a recent FBI bulletin says otherwise).
- Baton Rouge: previously street gangs, now only drug gangs.
- Erie (Pennsylvania): four organized drug gangs, initiated from outside cities, including Detroit, with documented importation of juveniles to staff the crack houses. Erie is an exception in this test: not one of the one hundred largest cities but a clear example of the franchising process.
- Fort Wayne (Indiana): one street gang and three drug gangs initiated from outside (Detroit, Chicago, Indianapolis).
- Jackson (Mississippi): primarily drug activity but tied to residential territories and weak leadership.
- Jacksonville (Florida): adult drug gangs and small wannabe school groups slowly evolving into turf-based gangs.
- Jersey City: Asian extortion gangs (the infamous "Born to Kill," using Jersey City as a "national safe house") plus black drug gangs, but no street gangs.
- Lincoln (Nebraska): new drug sale recruiters from Omaha and Kansas City, but they "haven't gotten a foothold yet" among local youth.
- Lubbock (Texas): Hispanic traditional street gangs, with black gangs developing around drug sales.
- Oakland: some Hispanic turf gangs, more Asian gangs, and eleven black drug gangs, but no black street gangs.
- Oklahoma City: originally Hispanic, then Asian, and then black street gangs, some of which started around a drug focus and then became more criminally versatile.
- Shreveport (Louisiana): crack and gangs are "indistinguishable."

There is little pattern here or in the less interesting contexts that I haven't reported. Keeping in mind that drug or crack gangs are very much in the minority, it seems wisest not to draw conclusions yet about the outside and local forces that produce drug gangs. We do know enough, though, to attempt to differentiate between drug gangs and the broad array of groups that comprise street gangs.

Accordingly, I've outlined in Table 4-4 a number of ways in which they seem to differ. Remember that the list refers to dimensions, not absolute points, and so there is room for quite a bit of overlap.

There's also an interesting difference between street and drug gangs from a potential victim's viewpoint. You are safer from crime in a drug gang territory (of which there are fewer, in any case) than in a street gang territory. There is less violence and less property damage. The dealer is interested in you as a buyer. If you're not a buyer, you're not of much value (although when in need of money to buy more drugs for sale, drug gang members, like heavy users, are perfectly capable of burglary or robbery as a means of satisfying their need).

One critical point doesn't show up in this comparison. Given the proper urban conditions (I'll speak about this in Chapter 7), street gangs can evolve just about anywhere. In fact, they have, as the gang proliferation data suggest. But drug gangs can't (and crack gangs seem to be even more constrained), because they require an adequate market of users, a sufficiently uncontrolled neighborhood, and connections with at least mid-level distributors. The combination is not ubiquitous.

An interesting form of validation comes from the national migration survey in 1,100 cities. When the police were asked what intelligence they gathered on gang members migrating to their jurisdictions, they almost never cited intelligence on drug connections of any kind. To the working cop dealing with gang migrants, the drug issue seems remarkably unimportant, whereas Skolnick and various local and federal agencies suggest that it should be just the opposite.

Another logical rebuttal at the aggregate level comes from a simple piece of logic. If Skolnick and others are right about the crack-driven expansion in gang existence, then cities with predominantly black gangs should emerge at a significantly higher rate than cities with predominantly Hispanic gangs, because crack distribution is demonstrably far greater among blacks than Hispanics. Yet when we look at the growth in both black-predominant and Hispanic-predominant gang cities from 1985 on—the crack explosion period—we find no such difference in the data from our gang migration study on the hundreds of gang cities that emerged in 1985 and thereafter. Yes, there are more cities in which black gangs predominate, but they don't increase in numbers any faster than do those in which Hispanic gangs predominate or those with other ethnic gang characteristics.

From 228 long interviews with police officers across the nation, as a second component of the migration study, we determined the number of cities among the 228 that had drug gang migration only. The number was 17, or just 6 percent of the total. Eighty-two cities had both street gang and drug gang migration, which is an additional 36 percent. Drug gangs are out there but are still the exception, still not creating the bulk of gang member migration.

A local (for me) validation comes from the Los Angeles Sheriff's Department. Of 207 gang killings in the LASD's jurisdiction in 1991, only 5 concerned drug sales, and the rest were the result of gang-on-gang combat.[62]

In summary, then, our national gang problem is far more with street gangs than with drug gangs. The latter really are an extension of our national narcotics problem and need to be dealt with through drug policies. The street gang problem can be addressed in many areas without undue concern for the drug or crack complications. On the other hand, the emergence of drug gangs does raise the interesting question of whether they are the harbinger of a major change: Are they the wave of the future?

The New Entrepreneurial Gang

As much as forty years ago, New York City thought it saw an equation of street gangs with drug sales. The New York City Youth Board undertook a major study of the equation because "it has been alleged that street gangs are the centers of organized selling of narcotic drugs, that the gang membership itself is by and large a population which itself uses drugs heavily, and that they recruit (by various means of persuasion) users among hitherto non-using boys."[63]

Then, as now, tying gangs to drug sales was a popular story. Then, as now, the story was an exaggeration of a small but discernible pattern. A large number of gang members, the researchers found, were not users. Among that portion who were, more than half had sold drugs at one time or another (not regularly), and more than half of these had "stopped their selling activity altogether. In some clubs, there are no current pushers at all."[64]

Several factors understandably feed the notion well beyond reality; that is, they expand the acceptability of the drug and street gang connection. One is the public ignorance of both drugs and gangs. Another is the publicity given to particular instances of the connections that provide the basis for rampant generalizations. Early in the crack era, for example, five known Los Angeles Crips were arrested in Seattle for attempting to import and sell crack in that city. Other Crip sightings in Portland (Oregon), Anchorage, Phoenix, and Shreveport fed the perceived pattern: If there, why not everywhere; if a few, why not many?

Also, particular instances of gang members getting rich from distributing crack fuel the fire. Distributors were reported using Mercedes-Benz automobiles. Ricky Donnell Ross, known popularly as "Freeway Rick," ended up in an Ohio prison after a brief but glorious career during which he allegedly grossed $1 million to $2 million a day from controlling crack sales in Los Angeles and elsewhere.[65] His reputation rivaled that of Brian ("Waterhead Bo") Bennett whose real estate and other property purchases became legend. Keith ("Stone") Thomas became a folk hero in the Rollin '60s, perhaps the best-known large black gang in Los Angeles, one which included two identified cliques that concentrated on drug sales. Stone Thomas's legend seemed to grow after his partially decomposed body was found in a rented car out in a San Gabriel Valley avocado grove, handcuffed and shot in the head.[66]

So-called big winners like these—usually in prison or dead when we learn of them—do for local kids on drug streets what Wilt Chamberlain and Michael Jordan do for kids on the basketball court. Moreover, they feed the fantasies of the general public precisely because they are given so much credence as "typical" successes by the police and the media.

Yet another fantasy-feeding factor is the concern with conspiracy that is one of the hallmarks of federal enforcement agencies. Starting from the drug end of the connection, the Drug Enforcement Agency (DEA) necessarily concerns itself with conspiracy, with drug cartels and interstate trafficking. It is the most natural thing in the world to apply this same thinking to street gang involvement because in the

Table 4-4. Common Differences between Street Gangs and Drug Gangs

Street Gangs	Drug Gangs[a]
Versatile ("cafeteria-style") crime	Crime focused on drug business
Larger structures	Smaller structures
Less cohesive	More cohesive
Looser leadership	More centralized leadership
Ill-defined roles	Market-defined roles
Code of loyalty	Requirement of loyalty
Residential territories	Sales market territories
Members may sell drugs	Members do sell drugs
Intergang rivalries	Competition controlled
Younger, on average, but wider age range	Older, on average, but narrower age range

[a]With only one exception, every drug gang of which I've heard has been male. The exception is the autonomous female gang studied by Lauderback et al. in San Francisco, ("Sisters Are Doin' It for Themselves: A Black Female Gang in San Francisco," The Gang Journal, 1992).

absence of direct experience with street gangs in their own local set-tings, street gangs "sound" conspiratorial. The Bureau of Alcohol, Tobacco, and Firearms understandably falls into the same trap, as does the FBI. No wonder, then, that officials in the U.S. Department of Justice, fed by their agents from below, buy the conspiracy approach to street gangs and drugs, as may the department's branches such as the National Institute of Justice.

As much as I am concerned that these agencies are promulgating a warped image of street gangs, it is not without understanding how they get there. Conspirators certainly are entrepreneurs, but if the conspiracy to sell drugs is a relatively minor part of the total gang scene, then the "new entrepreneurialism" remains in doubt.

Still another source of the gang–drug story, of course, is some members of academia, only a few of them, fortunately, not most. I've suggested already that Skolnick's position seriously overstates the connection; nonetheless he has helped frame the problem with his concept of the instrumental or entrepreneurial gang.

In *The Gang as an American Enterprise*[67] (the title tells the story), Felix Padilla presents a case study of one Puerto Rican, drug-dealing gang in Chicago. It seems to be an offshoot of a larger street gang. As a case study that is never related to other gangs, street or drug, the work has unknown generalizability but it's a start, nonetheless.

The membership seems (Padilla never says) to involve about twenty-four males, most between 15 and 18 years of age, working for distributors in their twenties. They seem to be less criminally versatile than street gang members, more cohesive, more hierarchically orga-nized, and more responsible to the distributors. They are very focused on their dealing roles. In other words, Padilla's description nicely fits that offered in my comparison a few pages earlier. It's a drug gang, not a street gang.

This is a well-written book, but it does not attempt to place this group in an array of drug gangs—Padilla reports that there are a dozen others around—so we can't learn about the generic phenome-non. Nor does he compare them with their parent street gang or other street gangs. We are left to make such judgments ourselves. My guess is that Padilla's description, however, provides effective insights into the limited world of the entrepreneurial drug gang. To read it is also to understand why it is not a street gang.

I'd love to assign someone to apply Padilla's description to other purported drug gangs. For instance, the Pasadena (California) Police Department has acknowledged such a group as a small offshoot of one of its major traditional black street gangs.[68] Wouldn't it be nice to have an ethnographer available for such an assignment?

The two major new works that make the most deliberate case for the entrepreneurial gang are Carl Taylor's *Dangerous Society* and Martin Sanchez-Jankowski's *Islands in the Street*.

Taylor describes the gang situation in Detroit as yielding a sequential typology of gangs that change from "Scavengers" to "Territorial" to "Corporate" organizations. These types may be major reifications, as Taylor uses them. He equates territory with the drug business, and his corporate groups are the sort of gangs suggested by the DEA. Generally, as the book progresses, Taylor comes to equate gangs with drug distribution.

I have three classes of problems with Taylor's work. First is the bias that seems to result from the fact that Taylor was the director of a private security force and that his interviewers were his security officers. This is manifested, among other ways, in a set of policy recommendations that strike me as hard-core conservative and suppressive. I see no acknowledgment of the methodological bias.

Second, Taylor generalizes his findings uncritically to other urban settings, among which Detroit may be the most devastated. Just as the Los Angeles findings must be tentative until compared with others, so must the Detroit findings. The rust belt is nowhere so obvious as in Detroit.

Third, and most important, Taylor's methods and description of them leave me puzzled. The "corporate" gang he describes predated crack, a point he has missed. His historical accounts are based on some interviews and unspecified other sources. In research terms, his neighborhood interviews seem troublesome. Gang members' verbatim statements (members 12 years of age and up) are taken at face value, with no analysis or conceptualization. One can contrast these with a number of adult interview excerpts that are more useful because they refer directly to community history and development. There is little data analysis and no offense data, family data, numbers of gang members, turnover rates, ages, or percentage going into drug use or sales. Furthermore, there are few drug data—kinds of drugs, amounts, distribution systems, and so on. Such absences also lead to errors when referring to other places, such as Los Angeles (see his page 99). Finally, there is far too much reliance on media sources.

In short, it is impossible to assess the validity of Taylor's drug gang descriptions. Since they differ so dramatically from other descriptions, they would seem to require a level and a kind of documentation far beyond what is provided. The entrepreneurial gang may indeed be well and thriving in Detroit, but Taylor doesn't help us know in what numbers, with what seriousness, with what history, in what variations, or in what contrast with regular street gangs.

The last work to be discussed is Martin Sanchez-Jankowski's *Islands in the Streets*, a very creative new look at street gangs in three cities at various intervals over a ten-year period. The book has proved to be quite controversial.

Sanchez-Jankowski offers an important thesis, that street gangs have undergone a major change in orientation, becoming defiant collectivities with strong formal organizational properties. He sees them as highly structured rational planners and sophisticated units, carefully balancing in their activities their own predations with their positive contributions to local community residents and politicians. Gone is the impulsive, unhappy gang member; he is now a rational, calculating, truly entrepreneurial force.

Unfortunately, like Taylor, Jankowski does not share his data with the reader. He gives us little description of them and no approach to their analysis. Since he explicitly rejects the pertinence of past research, it might have been useful to share the nature of the new research. As it stands, we have no way of judging the accuracy of his observations or the validity of the gang patterns he reports.

In summary, then, are there new entrepreneurial gangs? Yes, at least in the form of drug gangs, some have indeed emerged. We need to know more about them. Are they as extensive and marketwise as the media suggest and the public fears? Probably not, but poor or absent data don't adequately resolve the issue. At this point, there seems little documentation of this major change in the essential nature of the American street gang.

5

Handling the Problem: The "Softer" Approaches

The Beverly-Wilshire Junior Optimists Club, comprised of upwardly mobile "yuppie" white professionals and business men, was required to sponsor a youth group. They selected the Generals, a black street gang of over 100 youths from South Central Los Angeles. The Optimists' youth coordinator, noting that the Generals met weekly at the John Muir School, met with them with a slide show about John Muir and Yosemite Park. As he launched his pitch—"we could go camping, learn about the natural wonders of Yosemite and come closer to nature"—an influential Junior General from the back of the room moaned, "Aw sheeit, he ain't nothing' but a naicha lova." The Optimists soon thereafter terminated their sponsorship.

Klein, *Street Gangs and Street Workers*, p. 203

This chapter considers the various ways in which our society has responded to street gangs and their responses to these efforts. All the foregoing chapter materials—what has occurred in the street gang world, what has been learned about gangs—have been inextricably tied to our society's responses to gangs. These have run the gamut from denial to prevention to treatment to suppression. From the prewar days of Frederic Thrasher's time to the present, one might see the societal response as progressing from stage to stage as we accumulate experience in gang intervention.

I am inclined toward a different perspective, however. I think we are now entering the last stage of a single cycle, starting in community-level programming and circling around to that same perspective after some fifty or more years of ineffective activity. We have learned much about what not to do, though even this message has not been well communicated, but we have learned little about what we should be doing. Thus this chapter unfortunately has more negatives than positives.

Street gangs are no easy target; most attempts to influence them have brought more frustration than fulfillment. The rampant gang proliferation documented in Chapter 4, despite some major efforts to control street gangs, illustrates that we have not been successful. Let's briefly look at three approaches.

Gang prevention programs have been rare. They require accurate knowledge of the predictors of gang membership, that is, identifying likely future gang members, and they require knowledge of the causes of gang membership. Finally, they require knowledge of the likely impact of prevention efforts.

In regard to predictors, the best is an older sibling already in an established gang, but such a predictor yields a high rate of false positives. Residence in a gang neighborhood is often used, but not properly because most gang-age youth in gang neighborhoods do not become members.

The causes of gang membership are so broad, so interconnected that they are hard to isolate or treat separately. On the individual level they include a need for identity and status, along with some deficits in interpersonal relations. At the aggregate level, they include poverty, inadequate educational processes, population shifts, and ethnic segregation. We have been relatively unsuccessful in ameliorating these problems. It is worth noting, however, that they are not peculiarly American problems. Street gangs do exist in other countries, as I'll describe in Chapter 8.

In regard to the impact of prevention efforts, it would be wise to remember the importance of status and identity to potential gang members. Any prevention program that selects potential gang members and gives them special attention runs the risk of creating the problem it is aimed at preventing. Past programs in Chicago, Boston, and Los Angeles seem to have demonstrated this outcome.

Gang reform programs have been more common. They work with gang members to reform, rehabilitate, and divert them to more prosocial pursuits. They thus require knowledge of gang structure, of types of gangs, of the relevant environment, of gang behaviors, and of members' responses to reform efforts. By and large, these forms of knowledge have emerged after the programs rather than comprising the rationales for the programs. In the absence of such knowledge, reform programs have concentrated on group phenomena and, in some cases, have inadvertently increased gang cohesiveness and gang-related crime.

Gang suppression programs, most commonly run by police and prosecutors, have been targeted primarily at serious gang felonies and presumed gang leaders. They require the same knowledge as do the reform programs and make the assumptions of rationality and deterrence. The notion is that increasingly severe sanctions, applied more broadly along with the active harassment of gang members, will result in rational decisions by gang members to desist and by potential gang members not to join. The assumptions, however, are weak, and the ability to mount consistent suppression efforts have proved even weaker. Suppression programs have shown no evidence of success.

In sum, we have three approaches that are based on different assumptions and that have goals difficult to achieve and procedures wrapped more in ideology than in empirical knowledge. The challenge is for the preventers, reformers, and suppressors to use the available data to design intelligent programs with an empirical base.

Consider California, more affected by street gangs than any other state is, by far. According to the gang migration survey, the state has 196 cities with street gangs, 60 in Los Angeles County alone. The state's Office of Criminal Justice Planning in fiscal year 1990–91 poured almost $6 million into sixty projects under its Gang Violence Suppression Program. Included were school programs, street work programs, community mobilization, diversion alternatives, and a wide variety of criminal justice enforcement projects. Yet not a dollar went to an independent evaluation of the effectiveness of these projects. Sixty wasted opportunities to assess our efforts seems to be an inexcusable exercise in public irresponsibility.

Yet despite the paucity of respectable evaluations of gang intervention programs, we need to attempt a synthesis of what we think we have learned. Although we could slice the intervention pie into a variety of patterns, I'll present in this chapter a broad set of categories, including community organization approaches of past decades, street work programs, and recent reinventions of community organization approaches—"what goes around, comes around." In addition, I'll report on some very special creative examples of individual efforts to make a difference.

Community Organization Programs

Gang programs operate on a continuum of program goals, running left to right (from liberal to conservative, as it were) from gang "value transformation" (reform) to gang behavior suppression. The earlier reform programs, whether emphasizing community mobilization, street work, or provision of opportunities for youth, were designed to change gang members' values, attitudes, and perceptions and to resocialize them into acceptable youth. Behavior change would follow.

In contrast, current programs attempt to suppress gang formation, involvement, and criminal behavior by means of close scrutiny, vigilant police practices, and directed prosecution. There is little concern here with gang members' values and usually little with prevention goals. Keeping in mind this value transformation to behavior suppression continuum may help the reader follow the cycle of intervention program categories of this chapter, starting in the community organization pattern. In doing so, we should keep in mind George Knox's comment: "Sadly, in some of our American neighborhoods, the gang seems to function much more effectively and meet more regularly than do local community organizations."[1]

The prototype of community organization approaches was the Chicago Area Projects initiated by Clifford Shaw and conceptually bolstered by the sociological work of what became known as the Chicago school of sociology. Because gangs arose in disorganized and transitional inner-city areas, it stood to reason that renewing the social organization would lessen the functional need for gang structures. In particular, it was felt that local community groups—indigenous groups—should be socially and politically empowered to improve neighborhood conditions. Tying informal community structures to formal agencies—schools, enforcement, welfare—would provide the social structure for healthy socialization and vitiate the need for gangs and other forms of deviance. A proud community with the will and resources to handle its problems was the goal; it was assumed that gang reduction would be one natural consequence.

The Chicago Area Projects relied on several assumptions directly related to our interests. One of these was that much delinquency, the principal target of the program, was group related. The groups of concern ranged from play groups to street gangs.[2]

A second assumption was that the source of high delinquency (and therefore gang activity) lay in the residential patterns and community structures of the inner city. Ethnic successions and out-migrations yielded relatively disorganized communities that permitted the formation of gangs. The "fault" lay in aspects of inadequate community organization, not in individual deficiencies or specific ethnic cultural patterns.

A third assumption, logically following from the first and second, was that the reduction in delinquency would proceed from the development of cohesive community organization, especially one empowering local community residents to define their own local problems and determine their own solutions. In this latter, they could enlist the formal institutional structures of the city, for example, its recreational, school, police, social, and political agencies, but the solutions to the problems were to be developed by residents, not imposed on them.[3]

Different writers have stressed different elements of the Area Projects, but we can capture the uniqueness of the approach as follows:

1. The operational unit is the neighborhood, not the individual or group.
2. Although professional workers may activate the community, the planning and decision making is not in their hands; it is in the hands of local residents, and credit is to be accorded to them.
3. Local workers initiate community meetings, seeking as much participation as possible.
4. A local committee is established as the focal point of planning and policy. Literally hundreds of such committees were formed, representing the varied interests of local residents, usually with youth welfare and delinquency reduction as the main concerns.
5. Local workers—community organizers, agency staff, gang workers—become the daily force for program implementation, rather than public officials or bureaucratic agencies.

Whereas the various committees emphasized different concerns specific to their neighborhoods, the most common concerns were recreational facilities, campaigns for community improvement and amplified services, development of youth agencies, and direct intervention in delinquent behavior and gang activity.

Individual community or gang workers were perhaps the most creative single program component. Indeed, the concept of "detached workers" or "gang workers" developed by the Area Projects became the more narrow essence of major gang projects for decades to come.

How effective were the Chicago Area Projects? With respect to delinquency reduction or gang control, no one knows. Program evaluation techniques were not well developed at the time. The best-known comment on effectiveness, Solomon Kobrin's twenty-five-year assessment could conclude only that "in all probability" delinquency was reduced. This conclusion was based on the description of widespread program implementation and the assumption that "these achievements have reduced delinquency in the program areas, as any substantial improvement in the social climate of a community must."[4] This is a weak straw indeed on which to lean, based on sociological faith in the absence of empirical fact and belied by the assessments of the detached worker projects of following decades.

Beyond the Chicago developments, probably the best-known single experiment in community organization was Mobilization for Youth (MFY), begun on New York's East Side in the early 1960s by the combined forces of the federal and city governments and the Ford Foundation. This comprehensive effort was the embodiment of what later became the rationale as well for the community action programs of the Johnson years (see Chapter 3).

MFY was remarkable for three reasons. First, it was carefully planned on the basis of a well-articulated social science theory, Cloward and Ohlin's theory of opportunity structure. Second, it was a large social reform experiment that clearly exposed the resistance to change typical of established social and political institutions. Third, it was for the most part a controversial and massive failure to achieve lasting reform. The interested reader should consult Daniel Moynihan's insightful treatise on the experiment.[5]

In MFY, the gang intervention activities were merely one component of a complex web of intervention programs—educational reform, job training and a Youth Services Corps, neighborhood councils, youth centers, social service centers, and political involvement. Unfortunately, as was so often the case with community organization attempts, no serious attention was given to assessing the effects of the gang program, mainly a detached worker operation similar to those to be described on the next pages.

Still, for our purposes, it is important to have demonstrated that concentrated, targeted community organization programs can be undertaken, for in the 1990s they are being tried again. A pamphlet written about the original Area Projects by Helen Witmer noted three implementation outcomes that are pertinent:

1. Local residents can organize themselves with effective mechanisms for dealing with youth problems.
2. Such organizations can endure over long periods of time.
3. Local talent can be discovered and enlisted in the battle. One need not be dependent on existing bureaucratic entities.

Gang Worker Programs

Of all the forms of gang intervention, gang worker programs have taught us the most about gangs, gang process, and gang response to programming. As I noted earlier, one must particularly keep in mind the concept of gang cohesiveness because it contains the single best clue to the potential success or failure of any gang control approach. Cohesiveness at work is best seen in the operations of gang worker programs.

The New York City Youth Board

The community worker component of the Chicago Area Projects was picked up by the New York City Youth Board and other agencies in the late 1940s and expanded dramatically in the 1950s, a decade that saw one hundred gang murders throughout the city.[6] The board started in 1947 by funding the gang work of other agencies, but it soon became dissatisfied with their inadequacies and began its own program in 1950. It started in South Brooklyn and rapidly expanded to Manhattan, the Bronx, and Queens. More than 150 "conflict gangs" were identified on the Lower East Side, Chelsea, Park West, Central Harlem, and East Harlem in Manhattan; Bedford-Stuyvesant, Williamsburg, Redhook, and Brownsville in Brooklyn; Morrisania-Belmont and Mott Haven–Longwood in the Bronx; and Long Island City and Astoria in Queens.

To meet the challenge, the Youth Board expanded its three youth gang units with 3 supervisors and 19 workers to ten units with 10 supervisors and 80 workers.[7] By 1961, the number of workers had grown to 150. As street work programs go, this was a truly large-scale effort. Significantly, this was a city-run program, with some advisory committees made up of agency professionals and businessmen. The indigenous community input and control that characterized the Chicago Area Projects was nowhere to be seen. Community organization was a minor aspect of the Youth Board operation. Nor, for that matter, was it to be found in any of the street work programs to be described in this chapter. Like some good wines, Chicago does not travel well.

Even though the Youth Board did not undertake useful research designed to assess the effects of its operations with gangs, it probably did more than any other operation to define the role of the gang worker (or *detached worker* in the Youth Board's terminology). Certainly the role needed careful definition. As described by Robert Rice in a *New Yorker* "Profile," it is unclear why anyone would seek such work:

> By the criteria of an ordinary job-hunter, street-club work is totally without attractions. It is uncomfortable. It is depressing. It is lonely. It is exhausting. It is nerve-wracking. It is often frustrating, because a worker's first victories with a gang tend to be negative—the prevention of shootings, stabbings, beatings, and stompings that he can't know for sure would have taken place without him.[8]

The Youth Board's gang program did not solve the problems inherent in street gang work. Rather, it was an experiment in intervention, with ambiguities regarding its own goals and some confusion between goals and procedures. It lost the community focus and developed instead a rather narrowly focused worker program which, nonetheless, became the model for future street work programs.[9]

The first requirement was carefully selecting the workers. The Youth Board received five hundred applications and interviewed two hundred candidates to hire its first eleven workers. This is a selective standard probably never duplicated in any other such program. Workers were to offer "sympathy, acceptance, affection and understanding" to gang members—contrast this with the police gang suppression programs of the 1980s—in order to modify the antisocial stance of the gang.[10] The coin of the realm was the worker–member relationship, the rapport that would permit the gang member's value transformation.

Just about any activity held potential for enhancing this value transformation: club activities, athletic teams, and fund-raisers such as car washes, dances, trips, and parties. Note the heavy emphasis here on group activities, for these were later revealed to be cohesion enhancers rather than value transformers. That is, they made gangs larger and stronger.

In addition, the worker was encouraged by the Youth Board to try "passing around cigarettes, paying for pool, visiting boys in jail and accompanying them to court"[11] to gain acceptance. Then workers were to help by intervening in family or school situations, developing job-training and employment contacts for the gang members, providing individual and group counseling (and soliciting professional therapeutic help when necessary), and giving referrals to other agencies as appropriate.

Most of this was to be done in the gang's own environs—the streets, club houses, playgrounds, and hangouts of the gang—not in the worker's office. Thus the term *detached worker* was apt; he was detached from his desk and office and assigned to the world of the gang: street gangs and street workers.

The Roxbury Project

Detached workers soon appeared in a number of gang-involved cities—El Paso, San Antonio, Seattle, San Francisco, Los Angeles, Chicago, and Boston. But in most programs, the absence of research prevented any growth in our understanding of street gangs and how gang workers could most effectively respond to them. For instance, the Youth for Service program initiated in 1957 in San Francisco was a detached worker program with a very distinctive feature: Gang members were put to work in service projects throughout the city, such as cleanup and repair projects.

Such community service involvement by gang members might be expected to provide some job skills, commitment to community, sense of personal accountability and responsibility, and so on. But we'll never know because no research attended the program. The best that could be learned from an after-the-fact assessment by the Survey Research Center at the University of California at Berkeley was that whatever might have been the results with respect to gang behavior, its import was "its character as an experimental agency in facing the problems of socially handicapped youth and inter-club conflicts."[12] That's not much help.

Contrast this with Walter Miller's mid-1950s Roxbury Project. Here, in an inner-city area of Boston, voluminous data were carefully collected on gang structures, gang members, and gang behaviors. Similar data efforts in Chicago and Los Angeles later revealed that one could generalize widely on these matters from city to city across the country.

The Roxbury results were most discouraging. Over a three-year period, gangs assigned to workers fared no better than did "untreated" gangs with respect to delinquent behaviors, arrests, or court appearances. In fact, the project gangs showed an increase in delinquency among males and younger members, especially with respect to the more serious forms of delinquency. Miller also was careful to note that the implementation of the detached worker operations, especially the establishment of strong worker–gang member relationships, was very high. One result was the recruitment of new, young gang members. The recruitment of new members, the increase in more serious forms of delinquency, and the correlation of these with effective program implementation are clear suggestions that the workers helped enhance the gangs' cohesiveness.

The Chicago Youth Development Project

With support by the Ford Foundation in the early 1960s, the Chicago Boys' Clubs embarked on a major detached worker project. Directed principally by Hans Mattick, the Chicago Youth Development Project (CYDP) excelled at implementing programs and collecting data to demonstrate its full field operations over six years, starting in 1960. The CYDP lasted longer than the Roxbury Project, involved more gang members, placed more emphasis on community organization, and used ongoing data analysis to improve its field operations. If any detached worker project could prove itself, the CYDP was that project.

The workers effectively interacted with their gang members and had considerable success in moving them toward better family, school, and job involvement. But the same research showed these youngsters consistently backsliding as they approached the goals set for them by the workers. Value transformation, if indeed it were taking place, was not leading to useful behavioral change.

Furthermore, the rates of gang delinquency did not drop over the project period, and those gang members most involved in the project seemed to fare the worst. Consider the gap between effort and effect:

> Staff succeeded in finding 750 jobs for 490 young people; similarly, 950 school dropouts were returned to school 1,400 times. CYDP outreach workers made 1,250 appearances at police stations and courts on behalf of 800 youngsters.... Finally CYDP workers made 2,700 follow-up visits to

the homes of 2,000 juveniles who were arrested during the last thirty months of the project, in an effort to get them involved in one aspect or another of the project's programs.[13]

At first blush, it would appear to be "obvious" that such a busy project as the CYDP must have had some positive effects on these rates, but the fact is that it did not. Despite the successful efforts of the staff in finding jobs, returning school dropouts, and intervening in formal legal processes, the youth unemployment rate remained at about the same level; the school dropout rate grew slightly; and the arrest rates of youngsters in the CYDP areas increased over time, with a lesser proportion of them being disposed of as station adjustments.[14]

The Ladino Hills Project

Toward the end of Chapter 2, I illustrated the pivotal importance of gang cohesiveness by describing various aspects of the Group Guidance Project in Los Angeles. I don't need, therefore, to repeat the findings of that project here, other than to remind the reader of the tenor of those results: Increased group programming leads to increased cohesiveness (both gang growth and gang "tightness"), and increased cohesiveness leads to increased gang crime.

The Ladino Hills Project was a direct response to those results. We wanted to test these cohesiveness notions, in as controlled a fashion as possible, in a small pilot demonstration project. We weren't gang busting, but we were convinced from past research that many sources of gang strength were external to the gang—rival gangs, police pressures, and the alienation of schools, parks, and businessmen. If we could reduce the pressures from outside and selectively invene with gang members on the basis of their gang relationships (rather than their individual psychological needs), we could reduce their cohesiveness. We saw the gang as a caricature of normal groups, and as such it wasn't necessarily a "normal" or "natural" or "necessary" structure.

Moreover, we felt that gang membership was destructive to its members. It unnecessarily propelled them into the justice system and lessened their opportunities to benefit from school, employment, and other resources for normal development. Although the gang gave its members the status and identity they sought, it also brought deficits that could well damage their futures. Beyond this, of course, was the harm done by gangs to their victims and communities.

If we could make the gangs less cohesive, their members would be freer to take advantage of alternative resources. If the gangs became less cohesive, so would their delinquency, according to earlier research. We located the project in Ladino Hills with a gang called the Latins (both pseudonyms), made up of more than one hundred males and females with the highest rate of commitment to correctional facilities of any gang in Los Angeles County at that time (mid-1960s). This was no Mickey Mouse group of corner boys but a traditional gang with a thirty-year history of self-regeneration.

We accepted three tests of ourselves: (1) Could we successfully implement a program based on notions of reducing gang cohesiveness? (2) Could we actually achieve measurable reductions in cohesiveness? and (3) Would such reductions in turn reduce delinquency? The sequence had to be in that order to test our approach; that is, we could not allow ourselves to try to reduce delinquency directly by means of counseling or deterrent practices. The results were as follows.

1. With some exceptions, we were relatively successful in implementing the program. Formal group meetings were curtailed and then stopped. More than one hundred jobs were found for Latin members (who kept them for periods of a few hours to over a year).[15] Local recreational agencies reopened their doors to individual members. Tutors volunteered to spend time with those gang members still in school. Openings in the Job Corps, Neighborhood Youth Corps, Marine Corps, and other training opportunities were used. Individual therapeutic interventions were arranged. The emphasis was always on nongroup interventions and the selection of youths as intervention targets on the basis that they would likely reduce cohesiveness.

2. Depending on the measure used, the Latins' cohesiveness was reduced by about 40 percent (not bad, considering that fully half of the members were siblings and first cousins). Most of this reduction occurred over the first six months, so it seemed quite dramatic at the time. After this period, there was no new recruitment into the Latins so that, with the slow attrition of members leaving the group, the group overall became smaller.

3. An accompanying reduction in delinquency (measured by recorded arrests) of 35 percent was achieved, which could be attributed primarily to fewer members, but the per-member arrest rates remained relatively constant.

4. The reductions in cohesiveness, delinquency, and gang recruitment continued throughout the eighteen-month project period and a postproject period of six months. Compared with the Group Guidance Project and also with the Boston and Chicago projects, we had demonstrated, albeit in a small and concentrated case study, that a conceptually based gang intervention stressing gang cohesiveness could significantly offset gang delinquency.

5. Several years later, the Latins regenerated themselves, and Ladino Hills reassumed its preproject, gang-ridden character.

In the absence of a sustained effort and in the context of a community that failed to follow up on the experimental interventions, a gang area continued to be a gang-producing area. We had affected the Latins, but we had not affected their community. The lesson is both obvious and important. Gangs are by-products of their communities: They cannot long be controlled by attacks on symptoms alone; community structure and capacity must also be targeted.[16]

CIN and CYGS

In Philadelphia, the Crisis Intervention Network (CIN) was developed in the mid-1970s as a highly modified detached worker program with strong gang surveillance overtones. Based on erroneous information, it was transplanted to Los Angeles in 1980 in an enormous operation called Community Youth Gang Services (CYGS).[17]

The CIN workers were more than detached; they also were mobile. Teams were sent out in radio-dispatched cars by a twenty-four-hour crisis hotline to areas of tension or gang activity. Working in close liaison with police, they attempted to defuse "hot" situations immediately following police intervention and to continue area surveillance after the event. CIN also stressed the development of parent groups in gang neighborhoods. Unlike earlier programs, workers were assigned to areas and not to specific gangs. This presumably would lessen their contribution to gang cohesiveness, but it would also decrease their knowledge of specific gangs and greatly reduce their attention to individual prevention and treatment efforts. CIN was a transitional program between the traditions of detached workers and the surveillance and suppression efforts of law enforcement described later.

CIN was quick to claim dramatic success for its efforts, as the Philadelphia gang homicide numbers dropped from a reported peak of forty-three in 1973 to almost zero six years later. But no independent assessment of effectiveness was ever undertaken. The dramatic drop in homicides in 1974 took place before CIN was in place and continued while CIN was starting as a limited pilot effort. In addition, there is evidence that the Philadelphia police were cracking down on gangs following the 1973 peak (the evidence includes a number of illegal crackdowns admitted to us by police officials) and that they altered their recording practices to label some gang-related homicides as narcotics-related homicides, thus lowering the gang homicide numbers by an unknown fraction.

Finally, others in Philadelphia suggested that gang activity was muted by the establishment of a number of local organizations. Philadelphia is a block and neighborhood city, with a number of block clubs, mothers' and grandmothers' groups, and the antigang House of Umoja (a residence-based gang treatment program), which was very active during the homicide reduction period. CIN's claims for success are hard to accept at face value, given these other factors. If CIN did contribute, then the failure of it and its funders in the city to provide a research assessment is our loss as well as theirs.

Still, the "hype" over CIN's success was enough to convince the Los Angeles County Board of Supervisors, and later the city council, to transplant the "Philadelphia plan," as it was labeled, to Los Angeles. Centrally dispatched teams in specially marked cars were available twenty-four hours a day to respond to gang flash points. Parent groups were instituted. And before the program had even initiated its first patrol area program, the county supervisor most responsible for the program's importation made a public claim for its success.

Los Angeles County's gang homicides peaked in the summer of 1980 and then dropped dramatically. CYGS had not yet even started. Its county version was begun late in 1980, and the city version not until a year later, by which time the number of homicides had already dropped from 351 toward the 200 mark. As so often happens, peak periods of activity spawn new reactions that postdate the "natural" reductions that would have taken place anyway. The social science and statistics literatures are replete with such instances. CIN and CYGS are merely additions from another category of exemplars.

Our prediction at the outset, as CYGS was implemented in the community-less areas of Los Angeles, was that it would evolve over time (maybe *devolve* is the better term) into a more traditional value transformation program, such as those described earlier. And so it did.

The emphasis on radio-dispatched crisis intervention slowly gave way to social intervention efforts, group programming for gang members, and truce meetings and outings for gang members (and a trip to Hawaii for cooperative gang leaders). But the "Philadelphia plan" couldn't make it in Los Angeles because community activation requires communities and crisis intervention requires professional intervention skills that are not a common commodity among gang workers, many of whom were formerly affiliated with gangs.

The Gang Violence Reduction Program

In the mid-1970s a new detached worker program was begun in East Los Angeles by the California Youth Authority (CYA), the state's youth corrections agency. As written, the CYA proposal read like a deliberate perversion of my conclusions in *Street Gangs and Street Workers*.

CYA emphasized the transformation of targeted gangs. Group events like sports activities and truce meetings were implemented. Older gang leaders were hired as "consultants" (i.e., gang workers).

The effort was aimed at reducing violence, especially intergang violence, rather than general delinquency, as in the projects described earlier. Thus, cohesion-building activities were combined to some (lesser) extent with a violence suppression approach, much as Philadelphia's CIN and the early CYGS attempted this compromise between value transformation and suppression.

When I saw the menu of cohesiveness-building activities, my expectations were that the project would fail. However, an evaluation process that contained both conceptually questionable and competent components suggested a limited success. Over a four-year period, seven targeted Hispanic gangs showed a slow but steady decline in homicides and intergang violence between and among themselves. On the other hand, the violence between the targeted gangs and others not involved in the project showed a far more mixed and variable picture.[18]

It seems feasible that the feud-mediation process could have had a positive effect on mutually targeted groups. I say *feasible* because the trend is visible in the data, but the numbers are small and the source capable of unintentional manipulation to favor the project (Sheriff's Department data in the early days of data collection development, when many definitional and recording processes were first being codified). The lack of positive effect on external gang rivalries may tell us something important about gang feud mediation: It doesn't generalize beyond the narrow boundaries of the targeted groups.

In any case, my worst fears were not realized. Further research attention to such intensive efforts as took place in this CYA project certainly seem warranted. But they should be guided more by conceptual underpinnings than this one was so that we can better understand why it succeeded or failed and thus improve our intervention efforts.

Feud Mediation

I certainly didn't invent the concern for gang cohesiveness in handling gang feuds, but feud mediation as just described is a tricky business. If done through formal truce meetings between gangs or in longer-term intergang "councils" of gang leaders, the potential for cohesion building is considerable. Over the years I watched a number of truce meetings in Los Angeles (some with TV news cameras rolling!) and gang leader councils. Not only did they not generally achieve their desired results, but they often legitimated the gang leadership, the exaggeration of rivalries, and the spread of rumors about gang conflicts.

Intergang dances and other events have led to full-scale free-for-alls (shades of *West Side Story*). Intergang sports events, if not heavily monitored, became the occasion for validating already existent rivalries. In times of intergang tensions, I've watched "club" meetings turn into war councils (some defused by detached workers and others not).

The truce meeting is merely the most visible example of group building at work in detached worker programs. Those programs revealed a set of paradoxes directly related to gang cohesiveness. To reach out is to involve youth in formal programs. The more one reaches out to them, the more they will respond and join up. To program is to solidify. Club meetings, sports activities, truce meetings, and all other formal programming only solidify the structure of the gang. To recognize is to legitimize. Gang members want nothing so much as the status and sense of identity that come with membership. Assigning them a worker, harassing them on the street corner, and formalizing leadership by means of position titles provide the ties that bind. The greatest source of a gang's cohesiveness is rivalry with other gangs. The second greatest source is how outsiders reinforce the group, however inadvertently, in the ways they respond to it. To gang members, the street corner, the corner of the park, the entrance to the ally are "theirs," but the cop claims the same territory and challenges for it. "Gi' me that corner" says the Chicago cop and in doing so recognizes the gang's claim as requiring his response.[19] The organizations feed off one another.

In the next chapter on gang suppression programs, I will make the case that these enforcement-oriented efforts are likely to have exactly the same deleterious effect as their value transformation counterparts did: more gang cohesiveness, more gangs, more gang crime. The mechanism for all this is the set of group processes, including the elements of cohesiveness, that derive from the gang's subculture. My case was reasonably well put by the earlier-cited City College summary of the New York City Youth Board project:

> Membership in any group often modifies, for better or for worse, the behavior of its individual members. An esprit de corps develops and group leadership stimulates some responsiveness of attitude in the whole group. The young people of street corner gangs, often the products of slums where they brush up against some of the worst elements of society, are restless and disgruntled, seeking the outlets, opportunities, and status they lack. Under such conditions the more aggressive or more reckless members are likely to take control, and the rest are drawn into modes of behavior from which they would otherwise have refrained.
>
> Thus problem groups become conflict gangs, and thus the conflict gang in turn becomes occupied with jealous claims of "turf" exclusiveness. Much of the violence in which conflict gangs indulge takes place only because its members have acquired attitudes and entered into relationships that as individuals they could not have entertained. The gang fosters new prides and new fears, new loyalties and new enmities. While it is the

family–neighborhood environment and psychological maladjustment that motivate boys to join conflict gangs, the more violent nature of their delinquency is very often stimulated by the gang itself. The conflict gang takes on the overtones of what has been called a delinquent subculture.[20]

Gang Suppression Programs

When I was a kid, the word *suppression* was a "bad" word. We associated it with the Nazis in Europe or the treatment of blacks in the South. It was permissible to suppress a laugh but not a group of people.

Now, in the context of law enforcement, *suppression* is a "good" word. It connotes proactive police work (as well as prosecution and corrections). Gang suppression means active intervention against gangs, called forth for reasons fully legitimized by state and federal agencies (which are described next). Gang suppression is asserted to be the "right" thing to do, and because it is "right," it requires little further justification and absolutely no research evaluation to assess its effects.

How did we come to this point, this dramatic departure from the intervention strategies so popular in earlier decades? The answers are several and certainly include the following:
1. The failure, or absence of demonstrable success, of the community organization and detached worker programs of the earlier years.
2. The proliferation of gangs to an ever-widening array of cities.
3. The perceived increase in gang violence and "random" targeting of innocent bystanders. I say perceived increase because until recently this had not been documented in any valid fashion.
4. The explosion of crack cocaine and its purported spread and franchising by street gangs.
5. The swing of the sociopolitical pendulum from liberal to conservative philosophies. It started with the "law and order" rhetoric of the Nixon campaigns, although that rhetoric was inconsistently applied in practice. It flowered in the Reagan and Bush eras. In the gang arena specifically, the suppression philosophy bordered increasingly on repression in both Los Angeles and Chicago, the two acknowledged beacons on law enforcement's approach to the rising tide of gang activity.

Historically, the increased recourse to gang suppression approaches followed the slow decline of gang worker programs and in turn preceded a renewed interest in community approaches. However, the suppression approach has flowered in so many forms that I will postpone its treatment to Chapter 6 and move here to a discussion of newer community and other "softer" approaches.

Community Organization—Again

There's a popular phrase these days: What goes around comes around. This chapter began with a description of the early approach to gang problems, community mobilization and organization. The Chicago Area Projects and Mobilization for Youth were cited as prominent examples. Now decades later, the emphasis on community approaches has resurfaced. Few have concerned themselves as yet with issues of success and failure, but the absence of measurable success for other approaches gives a certain legitimacy to the new trend toward community organization.

Will they work more effectively this time around? Perhaps; they have the advantage of being responses to a far more serious problem—more violence, more drugs, more fatalities, and all this in areas far more affected by the persistent and pervasive poverty of the urban underclass. In addition, the proliferation of street gangs in small, less fully urbanized areas offers the promise of access to greater community organizational skills. Public and private agencies work together better when they have similar perceptions of the severity of the problem to be faced, and the severity of today's gang problem fosters such similar perceptions.

Although I will briefly describe a number of the newer attempts, let me make clear that these still are the exceptions. Such approaches require strong leadership and the recognition that law enforcement alone can't do the job. "As long as gang cultures exist, we are chasing our tails. Law enforcement cannot break the cycle, only social improvements can break it."[21]

The newer trend is signaled in Los Angeles by such agencies as Sheriff Block's. In 1991 and on his own initiative, Sheriff Block organized a two-day workshop for more than two hundred community leaders. Undersheriff Robert Edmunds admitted that "our experience is ill-suited to preventing the emergence of new gangs or the increased membership of existing gangs.... Obviously, we miscalculated the solution.... What is needed are partnerships involving all segments of our society."[22]

When New York governor Mario Cuomo's New York Task Force on Juvenile Gangs completed its work, it too eschewed the enforcement-first approach, entitling its report "Reaffirming Prevention." Its recommendations were headlined on each paragraph with the words *prevention, prosocial, comprehensive, coordination, school-based, community-based, youth input, prescriptiveness, staff training, program integrity, program intensity, constructive gang functions,* and *evaluations* (an unusual but well-appreciated suggestion). Rightly or wrongly, there is no dominance here by gang suppression.

In considering the new flowering of community antigang programs, described next, the reader may want to keep in mind a distinction between the community being used by the police as an agent for gang crackdowns versus the community as an entity for absorbing its own gang problems. An example of the former is the Los Angeles project, in which the LAPD's CRASH (Community Resources Against Street Hoodlums) officers collaborate with probation officers concerning youths recently released from a term in probation's secure facilities. These returnees are referred to community resources, tracked and supervised, and subjected to police/probation sweeps. In contrast, the LAPD's former "gang czar," who never did feel comfortable with his department's extreme views on gang suppression, stated after settling in as the new chief in gang-ridden Colorado Springs, that gangs

> are not a law enforcement problem. Putting more kids in jail is not the answer to the gang problem any more than putting drug addicts in jail is the answer to the drug problem. There needs to be a strategy, a well-thought out, multi-disciplinary strategy involving all aspects of the community, not just law enforcement.[23]

This is in line with a Philadelphia task force proposal "that every effort made to solve the problem of gang violence should be, to a maximum degree, neighborhood-oriented, neighborhood-staffed, and neighborhood-directed."[24]

One of the perpetual problems, of course, is what might be called the practitioner's curse, the trained-in, peer-supported assumption that what we do, works. As one delinquency professional explained, "I can't believe that fifteen minutes with me won't help a kid." To accept the notion that we are individually ineffective or as professionals we are inappropriate is a very difficult pill to swallow. It's even hard to accept major modifications in our preferred approaches or to hear that we are properly seen as one small cog in a larger intervention enterprise.

Yet with the failure of our past and current operations, we would be well advised to put more energy into those larger enterprises. Two points are relevant. The first is simple: Street gangs are by-products of partially incapacitated communities. Until we dedicate the state and federal resources necessary to alter these community structures, gangs will continue to emerge despite value transformation, suppression, or other community efforts. I'm talking about the most obvious resources—jobs, better schools, social services, health programs, family support, training in community organization skills, and support for resident empowerment. That's easy to say but obviously not easy to do.

The second point, because it's more manageable, is even tougher because it's a challenge to which we can respond. I'm talking about establishing community-based coordinated programs to deal with the street gang problem and, I hasten to add, to design such programs so that they don't contribute to gangs' cohesiveness. Obviously I can't provide a blueprint for this—each community needs to design its own. But I can offer brief sketches of what some cities have attempted or are now attempting by way of comprehensive programs with a strong community organizational emphasis.[25]

- In Austin, Texas, whose mayor requested a task force report on gang control, the task force wisely altered the focus to a more general emphasis on "youth at risk," thus avoiding, at least initially, the problem of building gang identity and cohesiveness. Gang members are left to enforcement agencies.
- Similarly, Columbus, Ohio, developed a Youth Outreach Program that focused principally on prevention with at-risk or fringe gang members. Detached workers were employed not to become gang mentors but to connect with fringe members and refer them to alternative activities and agencies. Columbus included a strong school component.
- Fort Wayne, Indiana, formed the Coalition of Youth Services as well as a more specified Gang Prevention Action Group. Interagency coordination was combined with a generally deterrent approach, and less emphasis was placed on mobilizing community residents.
- This contrasts with the Community Reclamation Project in the Los Angeles Harbor Area, where federal funding through the probation department, with police coordination, was used specifically to mobilize residents, initiated graffiti removal and neighborhood watches, organized sports and other youth activities, started parenting programs, and, significantly, worked closely with the LAPD on civil abatement and STEP Act suppressions.[26]
- Also in the Los Angeles area, a joint effort by the police and the boys clubs in El Monte produced a more narrowly focused program that emphasized jobs for gang members, including their preparation for jobs. The most interesting aspect of the El Monte program was the degree to which the police became actively involved in this job arena. They were proactive interveners rather than limiting themselves to the more traditional crime-fighting role. Another unusual aspect in El Monte was the reliance on informal, personal networking rather than on a formal, interagency task force approach.
- In Aurora, a sprawling suburb of Denver, the response has been communitywide and multifaceted—but most interestingly, it

has been directed by private agencies with no governmental funds.

- There's the example of Tarrant County, where a task force approach included Dallas, Fort Worth, and a number of smaller surrounding communities. This program, recently developed, is perhaps the most comprehensive of those listed here. It spans prevention, gang intervention, and enforcement in its design, referred to as a "holistic" approach. There are ten subcommittees covering a host of concerns—schools, the media, and so on. Most interesting to me is the superior background work exhibited in the task force design, which carefully uses past research on gangs to inform the design and also carefully assesses the gang situation in the several communities addressed by this comprehensive program.

- In Evanston, Illinois, a community with its own indigenous gangs as well as imports from Chicago, the response was eventually spearheaded by the police. Program elements were added sequentially in community education, parent groups patrolling neighborhoods, recreation, youth organizations, and the like. The coordination between justice and community agencies was the principal hallmark of this development.

- In Wichita, Kansas, public and private agencies came together to develop a comprehensive plan before initiating specific activities. The plan outlined a series of undertakings under the broad heading of prevention and education, direct intervention, and suppression. The approach is comprehensive but without the structure found in some other communities that dictates priorities among the many undertakings.

- Finally, there is the much-copied Paramount, California, program. Stressing prevention, this program operates from a school base, offering a sophisticated repertoire of antigang curricular materials, ranging from coloring books for the youngest pupils to classroom demonstrations of gang resistance techniques for older youths and neighborhood meetings for parents. The emphasis is on targeted early prevention with materials readily adaptable to other communities and school districts. Notable among the several components of the educational program is the emphasis on teaching direct alternatives to gang participation.

These are not the only community programs in the country, but they appear to be fairly typical of a far larger number of attempts to respond to gang proliferation over the past dozen years. Their significance does not lie—not yet, at least—in their particular forms because the evaluations of effectiveness aren't available. Rather, their significance lies in both the reawakening of community approaches and the comprehensiveness of most of them.

Yet I fear it may be too late for the largest gang cities, like Los Angeles and Chicago, where uncontrolled forces will determine gang growth and decline. Smaller, emergent gang cities, however, like those just listed, may still be able to coalesce around common perceptions of both prevention and control needs. If state and federal governments get serious about reducing street gang problems, they might start with careful assessments of these new community endeavors, to ask how public resources might best be used to maximize their value.

Alternative Interventions

Lest the reader think it's necessary to be a gang cop or a community organizer to become engaged in gang prevention or control, I want to list a number of alternative approaches. They range from the weird to the wonderful. Although my examples come principally from Los Angeles, they are in only a few instances unique to that area. For instance, a staff member in our research institute is involved through her suburban church in a twelve-gang truce development, even though the church and its congregation are located in a decidedly nongang, upper-middle-class community.

In a similar spirit but a far different location, a park in South Central Los Angeles was the site for a massive demonstration—Turn the Tide—organized by some of the largest community organizations in the city. These agencies had enough clout to persuade other agencies, major local politicians, corporate sponsors, and media coverage to "walk hand-in-hand down graffiti-splashed streets," as the local paper described it.

Typical of activist church leaders in many inner-city neighborhoods throughout the county is Brother Modesto Leon in East Los Angeles. In addition to general antigang involvement and a special school system for gang members, Brother Modesto initiated the Concerned Mothers, an activist group that helps other parents, serves in the special schools, coordinates with other antigang groups, and patrols the streets in order to be in contact with (and against, when necessary) local gang members. This is, in some ways, the counterpart in sprawling Los Angeles to the smaller-area block groups and grandmothers' clubs of some of Philadelphia's gang areas.

Indigenous groups such as these may often offer more help than those initiated by well-meaning outsiders or public officials. The Urban Leadership Training (ULT) program in Philadelphia was specifically aimed at engaging gang leaders in employment training and opportunities, but the lack of follow-through so typical of gang projects left the targeted leaders even more discouraged than they were at the start.[27]

More creative yet even more out of touch was a proposal—fortunately not funded—by a private agency new to the world of gangs. The intent was to apply particular technologies of international diplomacy to urban street gangs. The simplistic parallel between the conflicting interests of nations and the rivalries of street gangs could in fact have been quite counterproductive, by yielding the kind of legitimization and reinforcement of gang leadership against which I have railed throughout much of this book. Creative activism in combination with academic gang expertise is far stronger than either by itself.

Creativity occasionally surfaces in the arts world as well. Not uncommon are attempts to turn graffiti writing into an art form with positive aims. Before the Los Angeles Olympic Games, for instance, professional artists were engaged to work with gang graffiti-writers to paint murals in the area around the major venues of the Olympics. Drama groups occasionally enroll gang members to create plays about their life predicaments. In one instance, a four-act play entitled *Gangs* was produced by a professional director after, it is claimed, auditions of more than 1,000 members of 130 Los Angeles area gangs.[28]

Moving further toward the strange end of the creative continuum, we come to a special educational tool developed by two conservative black probation officers. This is a board game, similar to Monopoly in design. Youth, parents, and community members are the players. Each move lands on a space, with the associated game card containing an "educational" message about gangs that affects one's place on the board. The message I happen to remember is "Gangs don't do no good for nobody," spoken dramatically by one of the officers. The board game was developed to teach the players about gang structure, symbols, crime patterns, and consequences. The message was preventive and suppressive, not rehabilitative: These two officers had no sympathy for the gang members on their caseloads.

One consequence of the explosion in gang-related assaults and deaths in Los Angeles County has been the engagement of the medical world (indeed, there have been gang shootings in several hospitals). Trauma surgeons at the county hospital have established seminars and written professional articles on gang injuries. The coroner's office offered its Gang Diversion Program. Along with gang member counseling, parent support and other referrals, the coroner's personnel propose to parade vulnerable gang members through the processing of bodies, including the observation of an autopsy (preferably of a fellow gang member). It is unclear whether this should be labeled prevention, treatment, or suppression.

Clearly in the prevention arena was the appearance of two older, fierce-acting homeboys, in full regalia of baggy pants, high-top shoes, Pendletons, and sunglasses, at a local elementary school where two groups of wannabe 10-year-olds had been threatening a gang-on-gang war. The two visitors threatened and challenged the two groups of fifth graders to the point of abject fear and denial of any true gang intentions. The two then stripped off their outer clothing to reveal their identity as police officers, called out for the charade by the concerned school principal.

Last in my list is the case of Father Gregory Boyle in East Los Angles, known by his gang moniker of "G-Dog" to the gang members of his parish. Greg Boyle is the most emotionally dedicated gang intervenor I've come across in many years. He is streetwise, empathic, energetic, and unflappable in the face of abject failure. He mothers and fathers his gang members. He is the ultimate street worker, a one-man agency with his own schooling and jobs program for "his" gangs. His one-to-one relationship with individual members is genuine, warm, and supportive. Greg Boyle epitomizes the very best in individual, caring gang intervention.

He also epitomizes the kind of ill-conceptualized reinforcement of gang life and cohesiveness that I have found to exist in almost every traditional gang city. Has his special jobs program at the parish really prepared his boys for adult employment outside the barrio? Have the hours of counseling really challenged and changed gang values? Is a long day and a heavy caseload the measure of success? It's surely better to reach gang members than not, but to what effect? As G-Dog says in his many inspirational talks in the community, he has buried twenty-six of his boys. But I have trouble seeing the translation of his heavy and loving involvement into the reduction of gang violence. There are so many ways to feed gang cohesiveness and so few to reduce it.

6

Gang Suppression

It is my philosophy that we attack gangs on two levels. We cut the leadership head off and remove them from the community while we beleaguer and attack the body and legs of the creature.

Director of one of the nation's largest police gang units, quoted in *C.Q. Researcher*, October 11, 1991, p. 760

In the mid-1980s, a senior official of the federal Office of Juvenile Justice and Delinquency Prevention (OJJDP) announced that the Los Angeles model of gang control had been selected as the prototype for gang programs nationally. Given what I knew of this "LA model" and the shape of the increasingly terrible gang homicide pattern in Los Angeles, I sought to dissuade the OJJDP from its plan.

My counterargument was three pronged. First, was it going to fund a set of programs nationwide based on a "model" associated with the largest and most rapid increase in street gang violence ever experienced? Second, had it read and considered the implications of research that seriously called into question the gang–crack sales connection that was used locally to justify the model? Third, had it seen any results in Los Angeles in terms of planning, program development, or actual coordination? Or alternatively, had the OJJDP (1) bought the LAPD's claims about suppression effects without seeking evidence, (2) ignored the presence of research about the gang–crack connection, and (3) overlooked the increasing levels of gang violence?

The result of our discussion was that the OJJDP's plan for gangs was put on hold as it was determined that the "LA model" was not a model at all but a collection of relatively uncoordinated suppression activities by different agencies, with unknown results. I have never regretted my intervention, especially because Los Angeles's gang homicide figures continued to rise year by year to its latest annual total of more than eight hundred.[1]

I'm not saying in all this that gang suppression should be abandoned. Indeed, I believe that we can't yet fully evaluate its utility. But to jump aboard its bandwagon without questioning that utility is silly and could in fact be destructive, just as faith in the value transformation model was determined to be. My plea is to look more carefully at existing suppression programs and the assumptions on which they are based (principally, the tenets of deterrence theory).[2]

What is the essence of the suppression approach?

1. Low priority is given to efforts at prevention and treatment. Enforcement officials often define these as being outside their mandate and expertise. In my survey of 261 police gang experts, only 96, or 37 percent, could name a nonenforcement person in the community who was familiar with gangs. Many of these were school security officers.

2. High priority is given to "street crime," a politically tinged concept that combines serious crimes against persons (assault, robbery, etc.) with crimes visible to the public (e.g., drug sales) and with crimes that frighten citizens (here we can add gang loitering, graffiti writing, and drive-by shootings).

3. A basic deterrence philosophy is adopted, accepting the notion that heavy surveillance, along with rapid, certain, and serious sanctioning, will lead gang members to reduce, or desist from, street crime. Thus the emphasis is on selective enforcement (less politely termed harassment), arrest, enhanced prosecution, conviction, maximum sentencing practices, and a correctional focus on probation and parole violations.

4. An underlying assumption of all this is that the targets of suppression, the gang members and potential gang members, will respond "rationally" to suppression efforts. Those caught up in the programs will learn from their mistakes (in addition to being taken off the street), and others will weigh the consequences of gang activity, redress the balance between cost and benefit, and withdraw from gang activity.

As the reader should be aware by now, a major concern of mine is that nowhere in all this is there any calculus of the gang members' perceptions, values, and entrapment in the social psychology of group process. Does the harassed homeboy merely shrink away from his corner in the barrio, or does he resent the police intrusion into his space and reinforce his ties to his homies? Successful deterrence does not inhere only in the actions of the suppressors but also in the reactions of the suppressed. It takes two to tango.[3]

Police Programs

The clearest embodiment of the gang suppression approach is in the programs developed recently by many—but by no means all—police departments. The spirit is captured by a police captain explaining his suburban department's approach:

> We do sweeps. We spend a day looking through their rooms, their houses. Those that we find in violation go to jail. Those that don't...are left with the thought that we're coming back. We'll announce when— when we knock at the door. Our attitude is that they're special, and we're going to treat them special—by putting them in the system and arresting them when we can.[4]

I asked each of the 261 police gang experts in my original survey whether their gang officers or units engaged in such functions as intelligence, investigation, suppression, or prevention (including community relations). Overwhelmingly, intelligence gathering was the most common function, with 215 responses. This was followed by crime investigation (98) and suppression (85). A mere 18 mentioned involvement in interagency or community task forces, and 21 mentioned prevention activities. In sum, with respect to the gang problem, police are generally involved in standard enforcement activity. The one unusual element in the picture is the introduction of deliberate gang suppression activities in one-third of the departments.

Asked to describe these activities, the respondents offered a dozen or so categories, of which special gang "sweeps" was the most common. This refers to a multiofficer dragnet through a neighborhood to detain as many gang members as can be legally tolerated. An example, Operation Hammer, will be described later.[5] Next in order came various forms of directed patrol or selective enforcement or harassment, sometimes described also as "hot spot targeting." In these procedures, known gang hangouts are given intensive surveillance. "Intensified patrol," said one respondent, "including bad busts, but it passes the message." Another described the department's approach as applying "excruciating pressure."

Other terms were given: saturation, special surveillance, zero tolerance, crime suppression units, tactical units, caravaning (several patrol cars cruising neighborhoods in tandem), and, in one instance, even a ninja-style unit, complete with black clothing. The message was the same—crackdowns in any form with which the department felt comfortable. These are not so much new programs as they are extensions, often military style, of already available tactics. Harassment, control, winning the "war against gangs" are both means and goals, with tactics seldom seriously related to any sophisticated understanding of the target groups. Success is not in the effect but in the doing. "We knocked the shit out of them," one unit commander told me.

Significantly, however, a number of my police respondents eschewed the suppression operations in some cases because they were not required and in others because they were distasteful. Street sweeps, in particular, were often specified as inappropriate. Because these were often tied to publicity about the LAPD's Operation Hammer, we'll take a closer look at this epitome of police suppression.

Operation Hammer was the LAPD's concerted effort at massive antigang street sweeps. In the spirit of former Police Chief Daryl Gates's militaristic approach, it quickly developed the characteristics of war games: war from the LAPD's perspective and games from the gang members' perspective.

Operation Hammer was a massive, antigang police sweep launched initially in the south central section of Los Angeles. A force of one thousand police officers was added to the area's normal complement on an April weekend in 1988. On a Friday night and again on Saturday, the force swept through the area, picking up everyone available on already-existing warrants, issued new traffic citations, arrested others for gang-related behaviors (flashing signs, wearing "colors," and so on), and arrested more for observed criminal activities (including curfew violations).

A total of 1,453 arrests resulted, requiring a special "mobile booking" operation set up next to the Memorial Coliseum. Arrestees were brought to the Coliseum, booked, and released for later adjudication. Of the 1,453 people arrested, 1,350 were released without charges.[6] Almost half were not gang members, and there were only 60 felony arrests, with charges filed in only 32 instances.[7]

This remarkably inefficient process was repeated many times, although with smaller forces—more typically one hundred or two hundred officers. Joan Moore cited a sequence of four weekend Hammers that yielded 563 arrests ("mostly on warrants," which can be served without an area sweep), three ounces of cocaine, two pounds of marijuana, and $9,000 in drug-related cash.[8] This amounts to an enormous waste of enforcement effort if these outcomes are used as the measure of effectiveness. Indeed, Morales reported that in one Operation Hammer episode involving two hundred officers, although the low yield of drugs and weapons was reported, there were in the same time period two gang-related homicides and three additional wounded victims.[9]

In San Diego, the police chief reported on a massive sweep by his new Special Enforcement Division: The first week produced 146 arrests, but mostly for minor offenses. At the end of the week, only seventeen persons were still in custody.

A veteran LAPD officer, a member of the first Hammer sweeps, described these to me as "all show—public relations." He noted that the sweeps were announced ahead of time to the media. Since Hammer took place on successive nights, the media coverage all but guaranteed that most gang members would remain off the streets, especially those most wanted by the police. Asked what effect the sweeps had on gang members, he responded that it probably gave them "good laughing material."[10]

The intent, of course, was quite the opposite. The Hammer was supposed to send a message, a deterrence message to present and potential gang members. The police, the message said, are in control here, not the gangs. But was this the message received? Let us ask two questions. First, how are gang members likely to respond to being caught up in the sweep? Second, to what extent does the sweep exemplify the known principles of effective deterrence?

In regard to the first question, imagine the situation of the typical gang member picked up in a sweep. He is contacted on the street or in his car, detained for a minor (probably traffic or curfew) violation, and whisked away to the booking trailer at the Coliseum. There he sees some of his homeboys, some rival gang members, and other arrested persons all being run through the booking operation as rapidly as possible so the officers can get back to the street operation. The scene is a bit chaotic, with little of the "majesty" of the law, and leads in a short time to our gang member's release.

Now he returns to his neighborhood and encounters or seeks out some of his homies. Does he say to them, "Oh, gracious, I've been arrested and subjected to deterrence; I'm going to give up my gang affiliation." Or does he say, "Shit, man, they're just jivin' us—can't hold us on any charges, and gotta let us go." Without hesitation, the gangbanger will turn the experience to his and the gang's advantage. Far from being deterred from membership or crime, his ties to the group will be strengthened when the members group together to make light of the whole affair and heap ridicule on the police. It does, indeed, become "good laughing material."[11]

What went wrong? Let us consider the few, most basic tenets of deterrence theory. For a punishment (arrest, processing, conviction, etc.) to be effective:

1. It should follow the comission of the illegal act as swiftly as possible. In Operation Hammer, booking was swift, but proximity to the act was not, in the case of outstanding warrants.
2. It should be a certain response; that is, punishment should result for as many illegal acts as possible. A Hammer arrest and booking responds to only one of scores of acts committed by the gang member and to only a few of the many members. The certainty principle is clearly not met here.
3. It should be severe in nature. A quick booking and release, usually followed by the failure to file charges, is quite the opposite; the sanction is mild.

4. Swiftness, certainty, and severity must be perceived as such by the sanctioned person. If the deterrent message is to be passed on to others, the sanction must be perceived as swift, certain, and severe by them as well. How the sanctioner (e.g., the LAPD) sees the message is not important. As I pointed out, I suspect that most gang members soon turned around any perceptions of deterrence. Group process won't allow an inefficient process like Hammer to threaten the gang but will pervert it to strengthen the gang, to accent the games part of the war games.

5. It should provide behaviors that are alternatives to the sanctioned behavior. Operation Hammer, as with all suppression programs, totally ignores facilitating desired behaviors. It seeks only to suppress the undesired.

If this analysis is valid, then Operation Hammer cannot succeed with gang members. Gang processes will disturb the intended message and even reverse the intended effect. Gang cohesiveness is easily fed by opposing forces, and justice officials are such opposing forces. But Hammer is merely an example of other suppression operations. It clarifies for us how they may build in their own failures because they do not represent carefully conceptualized interventions. They merely do more of what the suppressors already know how to do. Remember that one of the principal findings from the Chicago Youth Development Project was that when faced with the failure of their gang members to respond positively, the gang workers reacted by redoubling their efforts, causing even more failure to respond.

In the face of this lack of success, our first, knee-jerk response is to do yet more of what we have been doing. We assume that the fault lies with the level of effort, not its substance. Operation Hammer was the culmination of many years of frustration in the LAPD with the rise in gang membership and violence, despite every effort, such as the work of the CRASH units. It was merely a variation on doing more of the same.

The LAPD's CRASH units are a collection of localized units of uniformed patrol officers in standard patrol cars, the "black and whites" of TV fame.[12] CRASH officers are the LAPD's gang units, cruising the gang areas with the specific intent of providing heavy surveillance and harassment, along with gathering intelligence. David Freed explained the difference between CRASH and the LASD's counterpart OSS (Operation Safe Streets) in terms of suppression versus intelligence functions. He described a CRASH officer's nighttime surveillance: "The faces of the gang members are frozen in Fletcher's spotlight, their eyes defiant. `Damn, I'd like to jam those guys,' Fletcher says. He calls for back up, but other cars are occupied. `Damn,' Fletcher sighs, `I wish we could jam them.'"[13]

CRASH officers "jam" or harass and move on. "We've found that being friendly with these guys just doesn't work," reports another

CRASH officer, and the LAPD's war mentality is reinforced. In addition, CRASH officers typically rotate out after two or three years, never having the opportunity to become truly knowledgeable about their communities. This, too, is typical of the LAPD, as was demonstrated in the Christopher and Webster reports after the Rodney King beating and riot.

By contrast, the Sheriff Department's OSS officers are in plain clothes and plain (nonpatrol) cars as often as not and often remain assigned to a community for many years. They come to know their targeted gangs and members far better, and consequently they are in a better position to defuse hot situations, to anticipate violent and other criminal events, and to locate perpetrators and witnesses. Theirs is street-level intelligence, carefully nurtured.

Not that OSS officers can't be suppressive—they can, indeed, and often are. But the balance is different, the tactics more attuned to the nature of gangs, and the results may be considerably better: "Gang crime statistics provided by both the police and Sheriff's departments show that OSS—battling twice as many gang members with only a third as many officers—has done a more effective job in fighting gang violence than its more widely known counterpart, CRASH."[14]

Regardless of Freed's conclusions, the point is that special gang units can vary widely in the degree to which they adopt a suppression philosophy. It is the more suppressive versions that concern me, because I fear that they serve only to increase gang cohesiveness. During the postriot period in 1992, Crip and Blood factions continued a truce process that, according to LAPD statistics, seemed to be correlated with a reduction in gang violence (including homicides) in the central riot area. Nonetheless, the LAPD deployed a large "crime suppression task force" to the various truce meetings and the area in general. Having lost control in the riots, the LAPD seemed thereafter to be making up for lost time, regardless of attempts at community peace.

And of course, we can look beyond Los Angeles. Chicago police are notorious for some of their alleged activities. In 1981, a crackdown featured arrests for wearing gang sweaters and exhibiting other ganglike behaviors. The charge was "disorderly conduct" and led mostly to early releases and dropped charges. Chicago purportedly invented the process of apprehending a gang member, transporting him out of the area, and dropping him defenseless in a rival gang's territory. And Chicago had the officer nicknamed "Gloves," after his tactic of taking disruptive school students to the basement where he donned his protective gloves before beating them. Now there was suppression.

Chicago gang officers do for the Midwest what LAPD and LASD officers do for many cities, provide the suppression slide show. Milwaukee, for instance, was treated to "scary slide shows of murders and a display of gang weapons that would make the U.S. Army run for cover," with the message that without hard-line suppression, Milwaukee would soon resemble Chicago as a gang world. Oklahoma City got the same message from its police department.[15]

Boston police were accused by the state's attorney general of a substantial number of unconstitutional suppression actions: illegal stops, searches, and even strip searches, particularly of suspected gang members.[16] Columbus, Ohio, police were accused in 1991 of making unlawful stops and searches of anyone they wanted in the streets.[17] In Baltimore, investigative techniques are described as including interviews of "randomly arrested gang members" and having fringe members "placed in real or imagined jeopardy in order to bring them into highly stylized interview situations, designed to change the member's allegiance from the gang to the investigative team."[18]

A midwestern city invented a new technique. Police received permission from a local judge to give an arrested gang leader a choice, going to court and jail or going before a school assembly of more than 1,500 students to apologize for his violent act on campus, denounce his gang, and publicly shame himself. He chose the subtler shaming option.

In a western city beset by both Hispanic and Asian gangs, the police chief described to me his department's approach. The process involves an interagency enforcement team that targets the most serious 10 percent of gang members. The team uses vertical prosecution, civil abatement statutes, and "proactive" policing: "We'll follow the gang guy in our cars, let him offend, then bust him." When I raised the questions of legality and effect, the chief said, "Now don't go getting technical on me; I'm a wolf in sheep's clothing" to the legality query, and "I don't know what the effect is" to the second, "but I feel like some are slowing down and others are leaving town" (i.e., gang migration due to a local crackdown).

One other tactic should be mentioned because it is a combination of suppression and community responsiveness. Normally applied in the case of gang (or nongang) drug distribution, the "cul-de-sac" operation has received mixed reviews from local residents. A street, alley, or portion of a neighborhood where a drug market is flourishing is blockaded by physical barriers and police vehicles over an extended period of time. Residents may enter or leave (upon identification), but nonresident gang members, dealers, and cruising buyers are excluded from the cul-de-sac. Obviously, drug sales in the area drop precipitously, and many residents breathe a sigh of relief. But other residents resent what they see as a "siege mentality," and the dealers may simply set up shop elsewhere, in what criminologists call *displacement*. To my knowledge, there has been no independent assessment of cul-de-sac results, including displacement. Obviously, while the barricades are up, short-term efforts in the area are easily measured with crime statistics and do show the expected decrease in drug-related crime.

We can also look at the manner in which federal enforcement agencies have adopted suppression perspectives. I include here the FBI, the Drug Enforcement Agency (DEA), and the Bureau of Alcohol, Tobacco, and Firearms (ATF) in particular. One can also throw in some other elements of the U.S. Department of Justice.

The problem here is simple: (1) These agencies know and understand organized crime; (2) they do not know street gangs; and (3) they often assume the two are similar, when in fact they are not. The DEA in particular gets confused because drug distribution organizations such as the Colombian cartels and Jamaican posses share an interest in drugs with some street gangs. Calling each kind of group a *gang* leads to the application of cartel thinking to street gangs.

Item: The DEA describes the spread of crack cocaine from its U.S. origin in Los Angeles to cities north and east. Because in selected cases (Seattle, Portland, Shreveport, and others) Los Angeles street gang members (not gangs or gang units) are arrested for selling crack in these destinations, LA's Crips and Bloods are said to be franchising and controlling crack distribution. The latter, of course, does not follow from the former, but the DEA's habituated conspiracy orientation leads it to assume the equation. It then provides a rationale for the DEA to join the suppression effort.

Item: Attorney General William Barr (George Bush's last attorney general) assigned more one hundred FBI agents to street gang control, following the agents' release from cold war duty when the Soviet Union collapsed. These agents knew nothing about street gangs because they had had no experience with them. But they knew about conspiracies and federal interstate statutes such as RICO and other antiracketeering laws, and it was assumed that they could easily be transferred to the street gang arena. In Los Angeles the FBI agents were welcomed with open arms by the LAPD and also by the much-troubled Inglewood Police Department. They were scoffed at by the Los Angeles Sheriff's Department, the county's central repository of gang intelligence. It's a long way from Al Capone and John Gotti to the Crips and Bloods and barrios of East Los Angeles.

Item: The Bureau of Alcohol, Tobacco, and Firearms, whose intelligence reputation was hardly enhanced by the disastrous storming of the cult headquarters in Waco, Texas, and the death of four of its agents, entered the game by publishing gang manuals for law enforcement use. Based on secondhand information in most cases, the manuals describe specific gang structures, locations, styles, and symbols in cities throughout the nation. For a sense of the manuals' validity, I looked at the description of the traditional street gang prominent around my own university's area. The manual confuses north–south streets with east–west streets, undermining confidence in its other information. It reifies clique names and generic terms (e.g., homeboys) into structural entities. It places the gang as well in Utah, Idaho, Oklahoma, and Florida and then says "the boundaries...has [*sic*] now extended far beyond the Los Angeles area to Juarez, Mexico, and El Paso, Texas."

Item: Finally, I offer the example of a report to the U.S. attorney general from Lourdes Baird, U.S. attorney for the Central District of California. This report claims that Los Angeles's urban street gangs' activity is concentrated in drug trafficking (especially crack) and weapons-related crime. We know from previous chapters that this is wrong on both counts, but it does provide the usual rationale for suppressing gangs. The report also states that the Crips and Bloods "represent a genuine `organized crime' problem." The Justice Department (the FBI in particular) and the LAPD seem to have become quite "simpatico." The Los Angeles Sheriff's Department, however, has come more and more to view the gang/drug/conspiracy nexus with increasing skepticism. Its written response to Ms. Baird's report goes to some lengths to discount the organized crime image perpetuated by the federal agencies. It bears repeating that the sheriff's operation, though far from faultless, has, of all these agencies, consistently maintained the best intelligence on street gangs at the street level. The LAPD could have developed similar gang-level intelligence but for several organizational reasons has never chosen to do so. Being the larger and better-known agency, the LAPD's views are more commonly disseminated and accepted. The Feds clearly have chosen the LAPD's approach.

Schools and Suppression

Later in this chapter we'll see the growth of suppression programs in prosecution, in corrections, and in the framing of new legislation, but another example is provided by school systems across the country.[19] It should be understood that the proliferation of gangs has upset many school systems. The schools' generally placid existence has been increasingly threatened by students showing gang signs and dress patterns, by gang graffiti, by rumors of gang rivalries that could erupt into on-campus violence, and even by threats and assaults, with teachers as victims.

Schools have had to learn how to get into the enforcement business—security guards, metal detectors, seminars on gang culture, and the like. One of the most vexing problems turns out to be that of dress codes.

The reader might well assume that dress codes, being a form of gang suppression, would not find favor with me. In fact, however, I've been coming down on the other side of the issue. The school is perhaps the main place where potential gang members, from the most innocent to active "wannabes," come into contact with street gang culture. They can't walk away from it as they can on the street and in the park.

If gang members are forbidden to wear their jackets, colored bandannas and shoelaces, baggy pants or sports caps, and gang haircuts in school, they may drop out. This will not contribute to their future adult roles. Or they may continue to attend but with greater resentment, adding yet another source of gang cohesiveness.

These negative effects must be balanced against the potential benefits. Imitators and gang wannabes will find it harder to display the external accoutrements of gang life. Other students—the majority in most schools—will be even less tempted to try out these elements of gang culture.

Thus, although I have misgivings about antigang dress codes and would prefer more generic codes that don't target gang members only, I've argued in favor of dress codes during school hours. Recent legislation in Sacramento would authorize local school districts in California to require the use of school uniforms as a way of handling dress code issues. I don't know that this policy can properly be called suppression; behavior control is an alternative label. Gang dress and graffiti have become the side doors to gang membership: Practice the "art," and become accepted by the "professionals." To reduce gang fights on campus and reduce the viability of gang culture to the uninitiated, perhaps we should be willing to promote this form of behavior control at the expense of one area of free personal expression.

I thought I'd seen it all until I ran across the Youth Gang Unit of the Cleveland public schools. I imagine there are other school systems that by now have established similar units, but I admit to surprise at hearing from the Cleveland group. One of its members (there are five staff members) described the unit's functions to me as enforcement and investigation first, staff/parent/student training in gang matters second, and networking with other agencies third. The unit is located in the system's Division of Safety and Security, leaving no doubt about the tenor of its mission.

It is becoming clear that many schools are running scared. To whom do they turn? Very often, understandably, to the police and far less often, it seems, to social service agencies. And cops love to educate other professionals about their views of the gang problem. Gang identification procedures make for excellent training materials, as do the cops' "insider" stories about gang violence, the gang–drug connection, and the need to crack down before things get out of hand.

In concert with this emphasis on enforcement, federal agencies (principally the Office of Juvenile Justice and Delinquency Prevention) have established the National School Safety Center, located at Pepperdine University, in the Los Angeles area. The center has produced materials that, to me at least, represent the epitome of enforcement-oriented, misleading, violence-hyping material on gangs.

For example, the center produced a forty-two page pamphlet with a cover drawing of three gang members, complete with bandannas, dark glasses, and sneering smiles, carrying a club, a chain, and three firearms. "Gangs in Schools" offers as its very first written word beyond the title page, the following: "I never had a conscience. I tried not to let things bother me. You got to survive. If I didn't do it [kill people]...I'd get my head blown off" (former gang member).[20]

The pamphlet then lapses into more hyperbole and false information:

- "Larger gangs in major cities, however, often have members whose age ranges from 8 to 55."
- "Gang members today are much younger than those in years past."
- "Once in a gang, the odds are overwhelmingly against a boy (or girl) ever leaving it."
- "Large, well-established gangs frequently relocate from one city to another to establish themselves as the primary gang presence in town."

The hype includes inaccurate statements regarding the gangs' control of the drug trade. Public statements by the center's director go even further: "Fifty percent of those killed are bystanders."[21] This same writer, as recently as 1993, in the face of so much information to the contrary, commented, "Youth gangs, whose organization and existence at one time had primarily a social basis, now are motivated by violence, extortion, intimidation, and illegal trafficking in drugs and weapons."[22]

It is worrisome to see the occasional gang picture portrayed as the typical gang picture. The ideology of the past twelve years has led to an institutionalized distortion of street gang realities. Who can blame school systems for creating gang units and panicking at the threat of gang members in their schools if they turn first to the street professionals—local police—then to federally legitimated "experts"—like the National School Safety Center—for what they assume to be informed advice? The schools, after all, were never expected to solve the gang problem. Less distorting pictures of gangs in schools are available, and I commend them to the reader.[23]

Gang Suppression Through Prosecution

In the next few pages, I'll discuss suppression efforts by prosecutors, corrections agencies, and legislators. Together, they comprise an increasingly comprehensive package of crackdowns, with varying degrees of failure to implement the basic propositions of deterrence philosophy.

Prosecutor's offices—city attorneys, district attorneys, federal attorneys—are seldom concerned with street-level suppression. Rather, their issues have more to do with the evidentiary quality of the cases they can take to court and with encouraging as well as applying legal statutes that increase the number of "good" cases that they can prosecute. There are occasional efforts, however, that combine the statutory- and street-level forms of suppression, often through the medium of special court orders.

For instance, in the 1980s, the Los Angeles County district attorney asked for a court order declaring gangs to be a form of quasi-corporate structure, meaning that each member could be held accountable for the actions of the other members. The court acquiesced as part of the district attorney's plan to force gang members into a graffiti removal program. Gang members who may have had no part in placing this graffiti around town were nevertheless forced to participate in the program.

The DA clearly saw this development in terms of both specific and general deterrence. That is, the individual members forced to repaint walls would learn thereby not to be graffiti artists themselves, and other members would learn from these "dire" consequences of graffiti work that the practice was not worth the results. Was he right? Would the typical gang member be deterred by having to wash away someone else's artistry, or would he see another injustice here that confirmed his antiauthority predilections? Would removal of his gang's emblems shame him or merely confirm the importance of gang symbolism? Would his homeboys henceforth avoid graffiti work because they might have to wash some walls, or would they instead accept this as part of the moves and countermoves that characterize the oppositional culture of the street gang?

In the absence of certainty of sanction (the wrong members do the cleanup), in the absence of severity (wall washing is hardly a serious sanction), and in the absence of swiftness (weeks and months may pass between writing graffiti and removing it), we cannot be surprised if there is little deterrent effect here. And given the group dynamics that encourage gang members to turn every attack into a self-serving and group-serving validation of their status, we may even expect the DA's program to have a boomerang effect.

A similar fate may await an increasingly popular move by city and county attorneys who seek court approval of "gang-free" zones, usually local parks. In such instances, a park or a local neighborhood that has increasingly been taken over by gang members is carefully designated by the court as out of bounds to the gang. This court action permits heightened police surveillance and crackdowns, easier prosecution of misdemeanors and minor felonies that otherwise might have been overlooked, and a "time out" period in which local residents and authorities can restake their claim to the endangered territory.

Leaving aside the constitutional issues, I would ask again whether an ultimate outcome might not be reenforcement of the gang's status. There are other ways to combat gang takeovers; after all, harassment of citizens and other crimes can be prosecuted without publicly separating gang members from all other youths. Given the special attention and publicity, the gang's status is increased in its own eyes.

In this gang-free zone approach, the deterrence principles are weakly applied not only to past behavior but also to potential future behavior. For some members I have no doubt that future deterrence will be effective for a period of time, but for others it will present only a challenge. Further, some gang-free orders are so inclusive as to prohibit perfectly lawful behaviors, that is, lawful for all but gang members. Are the homeboys expected to shrink away, or will they feel even greater animosity toward a society that

> forbids lifelong friends, even brothers in the same family, from associating with one another (in the gang-free zone)...bans certain clothes and certain jewelry...bans...yet-to-be-named individuals from discussing or referring to the gang in any way, and forbids talk about any subject at all while riding in cars.
>
> The proposed order forbids each (member) from waiting for a bus or otherwise staying in any public place longer than five minutes.
>
> The proposed order would forbid 500 alleged gang members from engaging in the most ordinary lawful conduct on their own property and in their own homes. Children could not climb trees or fences.... Razor blades...baseball bats, flashlights, even screwdrivers would become contraband.
>
> Under the proposed order, 500 unnamed individuals would be required to carry special papers to prove that their everyday activities are lawful.

The quoted description is provided by Paul Hoffman and Mark Silverstein, two ACLU lawyers[24] and must be understood in the context of an impending legal suit. Even if the lawyers exaggerate, their description is relevant to my concerns. Once again, however inadvertently, the gang has been recognized and "legitimized" by our response to it. In this particular case, the Blythe Street Gang of the San Fernando Valley was named widely in print and on television. I wonder how many members have pinned the press clippings on the walls of bedrooms and hangouts or carry them in their jackets and wallets.

The Blythe Street order is somewhat extreme, and I use it as an example for that very reason, to illustrate the likelihood of a boomerang effect—short-term deterrent gain at the cost of longer-term increase in gang cohesiveness.[25]

Another extreme came in 1989 from the Los Angeles County district attorney. Declaring that LA gangs had become "a new and dangerous form of organized crime," the DA announced a new suppression program. "The objective is to use each occasion that a gang member is arrested for a crime, no matter how minor, as a means to remove him from the streets for as long as possible." In its coverage of the DA's announcement, the *Los Angeles Times* noted that a gang member arrested for drinking in public would thus be prosecuted with the goal of a six-month jail term. There is no pretense here about rehabilitation, assured the DA.[26]

The best-known prosecution antigang programs are modeled after, and often named after, the one initiated in Los Angeles by the district attorney. Operation Hardcore is designed to concentrate resources and knowledge in the prosecution of serious felony cases and cases involving known gang leaders. Its aim is to achieve the highest possible conviction rate in these targeted gang cases.

Central to the program is the use of "vertical prosecution," in which one deputy DA carries the case forward from investigation through conviction to provide maximum continuity (normally cases are handed off to several deputies on the way to final disposition). Other elements include

- Special training of police in unique aspects of gang warrants.
- Special training of the DA's investigators.
- Special training of police as expert witnesses.
- Witness protection programs.
- High bail requests.
- Elimination of plea bargaining.
- Emphasis in court on conspiracy and gang membership evidence.

According to an evaluation carried out in the early 1980s, Operation Hardcore in Los Angeles achieved an unprecedentedly high rate of convictions—95 percent—and understandably was copied widely across the country in response to the proliferation of gangs during the last dozen years.[27] Unfortunately, even as the Hardcore office in Los Angeles has grown from just a few deputies to more than forty, it has had to narrow its concerns to only selected gang homicides, as these have increased to more than eight hundred per year. What has emerged more clearly in the process is the distinction between specific and general deterrence as targets of the operation.

In regard to specific deterrence, suspects obviously are "deterred" during incarceration. Here, deterrence equals "incapacitation," and the only available victims become those inside the prison walls. Whether these felons remain deterred upon their release is almost a moot point: They've been convicted of homicides and will serve their terms regardless of whether prison is deterrent or criminogenic.

But in regard to general deterrence, it is far from clear that the successful prosecution of gang murderers will deter other potential gang members from murder or any other predations. More to the point, until very recently no effort has been made to attempt general deterrence by those distict attorneys' offices employing Hardcore practices. I'm reminded of the often-viewed practice in China of placing felons' names and pictures on posters prominently displayed on the streets of many Chinese cities. A picture with a red check mark (✓) denotes an executed prisoner, and there are many check marks.

The Chinese make it very clear that execution is meant as a deterrent message for everyone. If the district attorney wishes to increase general deterrence, he can't count on that happening as a "normal" consequence of getting individual court convictions. Instead, he must advertise his success and consequences; the certainty and severity of sanctions must be broadcast (swiftness in our court system is not possible).

In the last few years in Los Angeles, such a program has been initiated. Posters featuring convicted gang members have been distributed. The one I have in front of me as I write was issued jointly by the DA and the Pomona (California) police chief. It is headed "One less gang member in your neighborhood." The gang member's name and face are shown along with the charge against him and his sentence (in this case, a stabbing and a term of fifteen years to life). At the bottom is the message "Gangs—Prison! Think Twice."

Effective or not, such broadcasting of the consequences of gang activity is at least an attempt to improve the deterrent character of suppression programs. It could be improved by noting the high conviction rate, by using a background of many other "check-marked" pictures (certainty), and so on. But how many relevant people have seen the posters? They tend to be pinned up in the police department, probation and parole offices, and other spots unlikely to be seen by most gang members (or seen too late). The message is lost if not effectively delivered.

After all, arrests, court appearances, and court sanctions are part of the common core of gang conversation. In the unmonitored dynamics of a gang gathering, arrest and conviction are soon distorted by members to enhance their "rep" (reputation). Group respect, not disdain, often becomes the consequence of justice system processing. The poster becomes symbolic of the need to think through every societal response to gang crime, not just from the suppressors' and victims' viewpoints, but also from the perspective of the gang member as well. The message received must resemble the message sent.

The number and character of enhanced prosecution programs have grown with the greater seriousness and prevalence of street gang problems. A recent national survey of almost two hundred prosecutors' offices yielded the following list of statutory provisions now in use for gang prosecutions:[28]

- Transfers to adult court for juvenile gang members.
- Forfeiture of cars used in drive-by shootings.
- Enhanced penalties for crimes committed near schools.
- Confiscation of weapons.
- Assignment for damages to parents of gang-involved children.
- Enhanced penalties for graffiti writing.
- Prosecution for threats related to joining or leaving gangs.
- Prosecution for gang recruitment.
- Prosecution for criminal conspiracy under federal RICO and similar state laws in cases of drug sales and other applicable crimes.

Corrections Programs

In part because I have excluded prison gangs from my overall depiction of street gangs, I also have omitted here programs dealing with incarcerated gang members. But the field of corrections also includes probation and parole, and in some of these, gang suppression programs have evolved. Two in California present contrasting examples.

In Los Angeles, 1980 was thought at the time to have been the peak of gang homicides, at 351. In response, the LASD's gang unit was doubled in size, and the Community Youth Gang Services operation and the district attorney's Operation Hardcore were initiated. Also authorized in 1980 was the Los Angeles County Probation Department's Specialized Gang Supervision Program (SGSP). The reader may recall that Los Angeles's original detached worker program, Group Guidance, also was lodged in the probation department, starting in the 1940s and ending in the mid-1960s. The contrast between the two couldn't be much greater.

Group Guidance was a street intervention program stressing prevention: counseling, alternative activities, and the reintegration of gang members into the community. The gang supervision program, started in 1980 and still in operation, describes its purpose as follows: "Protection of the community is paramount. SGSP intends to accomplish this by removing hardcore offenders from the community, and by the use of custodial time for violations of probation."[29]

Ironically, the first and longtime director of this lock-'em-up suppression program, an ex-gang member from the 1940s, was once one of the Group Guidance workers. He personified a 180-degree turn, reflecting his department's substantial conservative realignment in the Reagan era. In the 1960s, this same man consistently ignored the carrying of vicious weapons by his charges—a machete in one case I remember, a chain and nail-studded club in another. Today, those weapons bring automatic referrals to court and return visits to probation camps or state institutions. It's not clear which approach is more meritorious, but the reversal in behavior points up the reversal in philosophy.

The SGSP targets a clientele broader than might be expected: (1) participants in gang killings and gang violence, (2) gang members who are self-identified or so identified by others, and (3) probationers who are "very likely to become involved in future gang activity." There are forty officers (and their supervisors) in the program, each having an intensified caseload of 50 (normal caseloads are 150 to 1,000 in today's funding situation). Although there is some prevention activity, the officers' main job is close surveillance, including unannounced home visits, street cruising, and arrests for technical violations and the most minor of offenses.

These gang probation officers are more community based than office based (note this parallel to the detached worker). Cruising with police and maintaining contact with deputies in Operation Hardcore are regular activities, to which school and other community contacts often take a backseat. An uninitiated graduate student of mine recently spent hours of cruising time with an SGSP officer and two LAPD patrol officers and was shocked at the degree to which the probation officer "outcopped" the cops in his confrontive, suppressive contacts with youth on the streets and in their homes. No pretense, she noted, of personal counseling or concern with family, relations, school, work, and the like; just warnings about attitude and behavior unbecoming to a penitent and quiescent probationer.

Is there good deterrence in this program? Like Operation Hardcore, it achieves some temporary specific deterrence by way of reincarceration. The heavy surveillance may also achieve temporary deterrence, for the duration of the youth's term of probation. But there hardly seem to be any built-in lessons for the future continuation of nongang activity and no serious attempt to build on alternatives. Severity is only a piece of the deterrence package.

In regard to general deterrence, the SGSP makes little if any effort to disseminate this message directly. Inadvertently, then, it leaves the message sending to the gang members themselves, and they are free to reject or distort it in their customary fashion. There are no data, of course, on the recidivism rates in the SGSP caseload or on changes in criminal activities among the probationers' homeboys. The program's success is measured in units of worker activity and probation violation rates.

A more ambitious program has recently been implemented in San Diego. The JUDGE program involves an integrated attack by police, prosecution, and probation, each employing agency-specific suppression tactics. The correctional (probation) component is described as follows.[30]

The JUDGE program stresses:

- Strict adherence to probation conditions, including special condi tions for gang members (curfews, not associating with other mem bers, not wearing colors) and drug testing.
- Accountability for the actions of targeted offenders with appropriate consequences for violations and new offenses.

Probation. The two probation officers in the JUDGE task force pro- vide information about targeted offenders to law enforcement and the dis- trict attorneys, and supervise a limited caseload of about 20 probationers. Most JUDGE probationers are actually assigned to probation officers in other probation supervision units. This highlights the need for coordina- tion with probation staff outside the JUDGE task force regarding actions taken on probation violations and new offenses. In July 1989, the Probation department initiated a Gang Suppression Unit (GSU) to provide intensive supervision for all high-risk gang members. GSU staff are housed in the same office as the JUDGE task force. Currently about 80 percent of the JUDGE caseload is assigned to GSU probation officers. According to JUDGE staff, this facilitates a coordinated response consistent with JUDGE goals and objectives.

When a targeted offender has a new offense or probation violation, the probation officer assigned to the case initiates court reports using information gathered from all components of the JUDGE program. JUDGE probation staff coordinate with prosecutors to ensure that project defendants remain in custody prior to trial or court hearing and the rec- ommendations for sentencing are consistent with the objectives for the JUDGE program.

Gang Suppression Through Legislation

Statutes outlining the nature of criminal and delinquent behavior, as well as the justice system's responses to these, are just as applicable to gang members as to the rest of us. For many decades of street gang existence, gang intervention got by with a modicum of special, gang- specific legislation. However, with the gang proliferation of the 1980s, the greater evidence of gang violence, and the hyperbole over gang–drug connections, things have changed, led as usual by California.[31]

Suppression-style programs have increasingly been backed up by, and in many cases initiated by, new gang legislation. A preliminary analysis by the Institute for Law and Justice in 1993 identified four- teen states that recently enacted gang-relevant laws and thirty-one states that adopted RICO-style laws, similar to the antiracketeering federal legislation that can be applied to gangs involved in drug traf- ficking or behavior that can be liberally construed as a form of orga- nized crime.[32]

Central to much of this legislation are two related rationales. The first is that gang membership in itself constitutes grounds for arrest, prosecution, and enhanced sentences. The second is that gang activity constitutes a form of conspiracy to commit crimes, and conspiracy brings with it additional laws and additional opportunities for sanctioning. Next I will describe several of the more common types of new antigang legislation, as well as some interesting variations.

STEP Acts

Late in 1986, I received a request from Michael Genelin, the newly appointed director of Operation Hardcore, to meet concerning some proposed new legislation. With a colleague, we sat down for a long lunch in our university restaurant to hear about the plans for a new deterrent approach to gang suppression, the proposed Street Terrorism Enforcement and Prevention Act (STEP). The words seem carefully chosen: *street terrorism* to emphasize the severity of gang activity and *prevention* to warm the hearts of liberal legislators. The proposal was creative, unabashedly suppressive in intent, and, to us, quite startling. It clearly bordered on unconstitutional repression rather than "mere" suppression and was based, in our view, on fallacious assumptions about gangs. Because STEP acts are now being enacted across the country, I shall quote the essence of the proposal as first drafted:

Section 423: Legislative Findings and Intent; Prohibited Street Gang Activity; Definitions

a. The legislature hereby recognizes that street gangs are involved in terrorism, and that the primary thrust of any street gang is criminal in nature. As defined, it is the intent of this statute to limit street violence and other criminal acts by street gangs through limitation of gang activity.

The legislature recognizes that it is the right of individuals to be protected from intimidation and physical harm caused by any activity of violent groups and individuals. Additionally, it is not the intent of this law, in any way, to interfere with the exercise of rights and privileges protected by the Constitution of the United States. Further, the legislature recognizes that it is the right of every citizen to express himself on any subject, and to associate with others who share the same or similar beliefs.

However, the legislature recognizes that violent activity directed against other individuals, or activity that is clearly criminal in nature is not constitutionally protected, presents a clear and present danger threatening public order and safety and should be subject to criminal sanctions.

b. Any person who becomes a member of, or maintains membership in a street gang with the knowledge that a purpose of the gang is to engage in assault with a deadly weapon, robbery, murder or sale or possession for sale of narcotics or controlled substances specified in section

11054, 11055, 11056, 11057, or 11058 of the Health and Safety code is by virtue of his membership in the street gang guilty of a felony.

As used in this section a "street gang" is a group of three or more individuals who associate together under a common name or common identifying sign or symbol and whose members engage in a pattern of criminal acts.

A "pattern of criminal" acts is defined as the commission of two or more of the above named offenses, provided that at least one of such acts occurred after the effective date of this statute and that any other predicate offense occurred within ten years of the charged offense.

It proved to be a most interesting, and most difficult, luncheon conversation. The Hardcore director pushed a conspiratorial view of the street gang with genuine surprise, I think, that we questioned it as vehemently as he stressed it. How could we, reputedly gang experts, argue with so many assumptions that, to him, seemed self-evident? As usual, part of the problem came from different exposures. Hardcore saw only the worst cases and never was involved in the normal, humdrum existence of most gang members on most days. Rather, this person saw violence as the core of gangs, whereas we saw it as a dramatic but overstated portion of gang life. Our differences were large enough, and the proposed law important enough, that I took the step—rare for me—of communicating my thoughts in a letter to the director for fear that our concerns might too easily fade in the mind of an advocate. I quote from that letter:

Dear Michael:

I enjoyed our lunch together and appreciated the opportunity to respond to your suggested legislation. I know you took a few notes, but here briefly is a compendium of my concerns:

1. Any time membership or association becomes felonious, I get scared. That's a knee-jerk reaction, and not specific to your legislative intent.

2. "Terrorism," as normally considered, does not apply to most street gangs or to most members. The terminology obfuscates rather than clarifies the nature of street gangs.

3. I dispute that "the primary thrust of any street gang is criminal in nature." Rather, the primary thrusts have more to do with identity, status, companionship, and perceived protection against perceived threat. Same comment applies about obfuscation.

4. Any person "who becomes a member" needs to be changed to "who is a member," remembering that joining may have preceded the current charge by a number of years.

5. The gang doesn't have purposes—members do. Don't overestimate the cohesiveness or stability of the gang structure.

6. Membership may be your slipperiest concept. It is, in the final analysis, judgmental. In fact, many youths claim membership which they don't really have.

7. Choice of "three or more" does not relate at all to the usual terminology nor to the objective reality of gang structure. Would you consider writing your legislation without reference to the term "street gang"? You could use "criminally involved group" and avoid most of my problems. Many criminally-involved groups are not "street gangs."

8. I remain, in any case, pessimistic about your hope that such legislation can act as a deterrent to either gang membership or criminal behavior, for reasons discussed at the lunch. Deterrence is far trickier than to be much affected by crack-down provisions.

I hasten to add that since that first encounter, Mike Genelin and I have worked collaboratively and with good rapport on several projects. Even more important, he was faced with the failure of suppression efforts to stem the gang tide and is now a strong advocate of enforcement/community coordination for gang prevention.

His original proposal ran into the expected resistance on constitutional and other grounds, was considerably modified both before and in the legislative hearings, but still emerged with its major intent quite intact. The essential elements include the following:

1. The term used is *criminal street gang*, to provide a less inclusive framework. The criminal street gang remains "three or more persons," no matter how informal the group structure, with one of its major activities being the commission of any of a specified set of criminal acts. This act is limited to serious offenses only, again narrowing the earlier focus. These are aggravated assault, robbery, homicide or manslaughter, drug trafficking, drive-by shootings, arson, witness/victim intimidation, and auto theft.

2. Rather than the all-inclusive membership proposal, the act defines its target as "any person who actively participates in any criminal street gang *with knowledge* that its members engage in or have engaged in a *pattern* of criminal gang activity [defined by the offenses noted], and who *willfully* promotes, furthers, or assists in any *felonious* criminal conduct by members of that gang." (I have put into italics certain phrases to point out the careful narrowing of the act's applicability.)

3. The "pattern" of criminal gang activity is further limited to involvement in at least two of the listed offenses.

These definitions provide some bulwarks against constitutional challenges. They exclude the bulk of gang-related crimes committed for individual rather than gang purposes—all the thefts, vandalism, minor assaults, weapons possession, drug possession for personal use, and so on. They exclude mere gang membership as an offense, and yet they allow enforcement agencies to concentrate on the serious crimes of most concern to them and on less serious crimes with probable gang goals.[33] So far, the STEP act has been copied and enacted in Florida, Georgia, Illinois, and Louisiana.

What the act also provides is enhanced punishments for convicted targets. Life imprisonment, if imposed, cannot mean parole during the first fifteen years. The court can impose up to three years of additional prison time for any convicted felony. Lesser enhancements are also available for jail terms.[34]

In deterrence terms, the STEP act adds a degree of certainty and some sanction severity as well. The added certainty of punishment is probably the more relevant deterrent property because of the way the act is carried out. To establish that any given street gang, and therefore its members, can be subject to the act, several activities are undertaken. In the new vernacular, a gang is "STEPped."

- Police and/or staff from the district attorney's or city attorney's office gather evidence from crime files and from interviews with local residents, businesspersons, and others that a targeted gang fits the "criminal street gang" definition.
- This information is then presented to the court, and the judge issues (if he or she is convinced) an enabling judicial order.
- Known gang members are then notified in writing that they are members of such a group. With such notice, the provisions of the act can then be applied to these members for future offenses.

The *C.Q. Researcher* journal cites one lighter moment in STEP act history: "In one instance, a gang member who received this letter from a district attorney crumpled it up and tossed it away, resulting in an arrest for the less-than-macho crime of littering."[35]

Is the deterrent message received as sent? A local precinct commander believes so: "From the hardcore to the wannabes, they know their next arrest will be dealt with severely."[36] In some communities, there have been reports that STEPped gangs have changed their names to invalidate the notices. An experienced ex-gang member, now an officer in the Specialized Gang Supervision Program, reported to me that even though the younger members don't understand the implications of being targeted, the older members make themselves scarce in order to avoid the official notification letter.

This notification procedure may indeed contain the seeds for useful deterrence. The "message" is literally passed from enforcement to gang member. The credibility of the message is enhanced, and legal implications of gang activity are clarified. Specific (member-by-member) deterrence is attempted, not just the ambiguous message sending of Operation Hardcore, with little of the boomerang effect that Operation Hammer seems to encourage by distorting the message.

Of course, misapplication of STEP can reduce credibility and therefore effectiveness. As occurred in Pasadena recently, gang members' homes were searched and members arrested in the belief that the act "makes it a crime to participate in a street gang."[37] The court later threw out the charges, and one might expect that these gang members and their homeboys felt vindicated more than deterred.

The idea of the written notification has been expanded in a number of states, usually through local versions of the STEP act. The definitions have not successfully been challenged in court, so there is good reason to expect further expansion of the approach.[38] Furthermore, we can expect embellishments. For instance, proposed legislation in California would have required gang members convicted under the act to be registered in much the same manner as habitual sex offenders are. My only wish at this point is that competent research could have been done in conjunction with these STEP acts. This has been our best opportunity to assess the deterrent capacity of a gang suppression program that actually uses at least some elements of deterrence theory. Once again, an opportunity to learn from our responses to gang activity has been lost.

Civil Abatement Laws

In the continuing search for laws to curb gang activity, various civil statutes have recently been invoked. Such laws allow us to deal with neighbors who party too loudly or allow junk and garbage to pile up. They also deal with failure to meet zoning restrictions or requirements for proper electrical and plumbing installation. Public nuisance laws fall under the same general heading.

This doesn't sound like the stuff of gang suppression, but consider the case of a gang hangout, be it a home or street corner. Or consider the case of a crack house or other drug-dealing location. Here the application of such laws not only to the gang members but also to the property owners, if followed by selective enforcement, could effectively suppress gang activity on the site, at least temporarily, and keep it "on the move" in a less organized fashion.

In its first major application in Los Angeles, the city attorney accumulated and presented to a superior court judge enough evidence to produce an injunction against the Playboy Gangster Crips, a street gang rather heavily involved in crack dealing. Under the injunction, the gang members could be arrested (and indeed were) and found in contempt of court for such acts as writing grafitti, blocking driveways and sidewalks, or any other form of public nuisance or intimidation of local residents. Other less obvious suggestions from the city attorney, such as leaving homes at night, gathering in public, or "boisterous behavior," were rejected by the judge as being unconstitutional.

All of this was made possible by the judge's ruling as requested by the city attorney, that the gang constituted "an unincorporated association," thus legitimizing inferences about gang organization and management and making any member party to an organized conspiracy. There is a parallel here to some aspects of the STEP act.

Not only are such laws used by enforcement officials, they also meet the needs of citizens. Local residents have taken landlords to court for allowing gang or drug activity to persist, with awards to the complaining citizens of $18,000 in Berkeley, $20,000 in Hollywood, and $30,000 in San Francisco. This time it's the landlords who are being deterred, deterred from permitting gang activity on their properties.

Landlords can be instructed, under the threat of fines, jail, or loss of the property, to improve lighting, install a security system, repair locks or fences, remove graffiti, tow abandoned vehicles, trim shrubbery, hire a manager, redraft the rental agreement, and so on. The Los Angeles city attorney has prepared educational material, *What You Need to Know About Gangs and Drugs: A Handbook for Property Owners in Los Angeles.*[39] The suppression message is spread to those who have the most to lose, the landlords. The gangs themselves are not the direct targets; rather, their activities may be suppressed through inconvenience or manipulation of their environment, much as is the case with the cul-de-sac operations mentioned earlier. In this sense, the suppression is more sophisticated in that it recognizes that street gangs have contexts and are affected by those contexts. Direct frontal attacks are not the only armament of enforcement agencies.

Closing the Public Parks

I mentioned earlier the somewhat Draconian attempt to suppress even the normal activities of the Blythe Street Gang in their general neighborhood. This was really an extension of a recent series of local ordinances targeting the use of public facilities. Local parks have traditionally been hangouts for street gangs, and sometimes they also are convenient sites for drug sales. Occasionally gang members come to think of a park as "theirs," and the general public comes to feel endangered there and excluded from using the facility.

Sometimes park managers can effectively run off the gang members, retaining its use for the 95 percent at the expense of the five. In Los Angeles, the issue was fully joined at Las Palmas Park in the San Fernando Valley. Using a city ordinance with a one-year limitation, the police were instructed to STEP the gang members with specific reference to use of the park, territory claimed by two rival gangs. The constitutionality of the ordinance has become more of an issue than its deterrent intent, and other cities are awaiting the outcome to see whether yet another weapon against gangs has become available. The ACLU has entered the fray to protect gang members' rights to free assembly. Meanwhile, gang members are avoiding the park and the written notices or arrests awaiting them there. The test of constitutionality has been passed in most cases like this but failed in one instance in Orange County, California. The city of Pomona added a second park to the action, and the county's Board of Supervisors has instructed park officials and the Sheriff's Department to compile a complete list of "gang-plagued" parks for possible inclusion in a much larger suppression effort. Most recently, the Los Angeles City Council proposed a 350-park ordinance.

In addition to the questions about deterrence is that concerning where the park-less gang members will go. Will they retire to their homes or to our streets? No other, more positive alternatives are connected to these suppression targets.

Additional Legislation

Perhaps the best window on new and creative suppression approaches is provided by federal racketeering laws (RICO) based on conspiracy assumptions and fueled by the war on drugs. Local recourse to federal statutes also allows direct collaboration with the FBI, DEA, and ATF. In the Los Angeles area, police departments have created relationships to the federal approach, ranging from outright rejection to total collaboration. Street gang involvement in drug distribution is a major determinant of the local–federal connection.[40]

The state of Wisconsin provides an example in the recently introduced Wisconsin Street Gang Crime Act. In addition to $1 million in grants to law enforcement and other antigang programs, the bill—known as Senator Chuck Chvala's "Gangbuster bill"—adds new penalties for such acts as drive-by shootings and gang recruitment. It also doubles the penalties for specifically gang-involved drug dealing, weapons violations, and battery. The senator's deterrent intent is explicit: "My bill is meant to send a simple message to youth gangs: Your time is up. We won't tolerate your reign of terror in the streets, schools, and neighborhoods of our cities."[41]

In California, the newest package of proposed antigang laws is even more diverse and includes

- Opening files and releasing the names of juveniles charged with serious offenses.
- Providing funds for a witness protection program (gangs are often noted for their attempts to intimidate witnesses).
- Permitting the tapping of phones and beepers of gang members.

- Prosecuting and sentencing 14- and 15-year-old murder suspects as adults (current law permits this at age 16).

Another proposed law in California, noted earlier, would require the registration of any convicted gang member, similar to the state law requiring the registration of sex offenders. Failure to register would itself be a misdemeanor.

In California and Minnesota, laws went on the books providing stiffer penalties for crack offenses than for any other drug, because of the gang–crack connection.[42]

In California, a late 1980s antigang legislation package included a provision for convicting parents who willfully fail to control their children's gang activities.[43]

Gang-related statutes, some proposed and some proposed and passed, set the unremitting tone of California's legislation in the late 1980s and early 1990s. Included were an easier waiver to adult court for certain gang-style offenses (drive-bys, witness intimidation, and so on), drug sentence enhancements, stiff penalties for drug sales on or near school grounds, prison terms for use of crack houses, broader use of seizure and forfeiture of property for drug trafficking (modeled after federal RICO statutes), sentence enhancements for use of firearms during drug offenses, new controls on automatic weapons, and special programs for graffiti removal.[44]

Finally, a flurry of new California laws passed in 1993 include a sentence to life in prison without possibility of parole for drive-by murders; authorization for school boards to prohibit gang-related clothing styles in school; and designating as a felony any attempt by an adult to recruit a minor as a gang member or to entice a minor into criminal gang activity. A bill before the legislature in early 1994 would allow a gang member, from the age of 12 up, who committed a serious felony to be remanded to adult court for prosecution.

As long as gangs continue to proliferate, we can expect that gang suppression legislation will also continue to proliferate. The only remarkable aspect of this trend is the continuing imbalance in the suppression/prevention proposals. Less remarkable, regrettably, is the absence of concern for assessing the effects of all this legislative effort.

As the reader knows by now, my guess is that the effects may, in many instances, be negative. I have made the point that the straightforward intentions of gang suppression programs may backfire. Their implicit deterrent properties and messages may be altered by the receiving gang members. Although this is less likely to be the case with nongang audiences, drug-dealing cliques may display the same character as that of traditional street gangs.

In time, we must survey the effects of suppression programs on both grouped and mass audiences, but for the moment I'm concerned specifically with the group setting. In transitional and inner-city areas where formal social controls are weak, group processes can more easily emerge to direct youthful behaviors.

Moore and Vigil suggest quite properly that "gangs maintain an oppositional, rather than a deviant subculture" and thus represent "an institutionalized rejection of the values of adult authority—especially as exhibited in the Anglo-dominated schools and the police department."[45] In a context in which a major law enforcement official declares that his department will "obliterate" gangs and that "casual drug users should be shot," this oppositional value system makes sense.

The gang literature is replete with descriptions of the oppositional, reinterpreting pattern of gang members, which were effectively restated in deterrence terms by Zimring and Hawkins:[46]

- "It seems possible that threats of punishment, so far from being disincentives to crime, may in these [gang] circumstances even function as incentives to it" (p. 216).
- "The operation of deterrence is greatly complicated when group pressures may not only inhibit the expression of the fear of sanctions but also in some instances convert stigmata into status symbols" (p. 317).

Thus the gang setting discourages the assignment of legitimacy to police, prosecution, and court accounts of acceptable behaviors. It denies the wrongfulness of many offense incidents (though not of all offenses per se). It encourages the bravado that accompanies antisocial deeds and utterances. It legitimates violence in the setting of gang rivalry and protection of drug dealing, and it accepts the gang's moral superiority in unequal battles and predations against both the weak and those in authority. In the area of drugs in particular, in which personal indulgence and profits are immediate and personalized, the credibility of antidrug messages is seriously endangered. And in the context of a neighborhood that tolerates trafficking, straightforward deterrence messages fall on deaf ears. The differences in effect between such programs as Operation Hammer and STEP acts could provide considerable illumination on how to use suppressive deterrence programming for control purposes.

7

Prospects for the Future

Graffiti no longer acceptable here. Please find a day job. Thank you.

Sign painted on a business building

So what are the prospects for the future? In the next decade or so, is the gang picture likely to improve? I fear not. First, our approaches to intervention and control are likely to be more of what we have already seen. Second, street gangs are the by-product of urban problems likely to increase in severity. Third, we've allowed—indeed, we could not have prevented—the widespread diffusion of street gang culture. Finally, there are signs that our unique American street gang is emerging in other nations as some of their urban situations come more to resemble ours. This chapter will be concerned with these issues. I apologize ahead of time for the pessimism, but I think it is simply realistic, just as it was ten years ago when John Quicker noted:

> In short, the general outlook for reductions in gangs and gang-related crime is poor. In fact, the opposite appears likely. Worsening social and economic conditions, federal roll backs of hard-won civil rights legislation and a potential increase in racism, and an expanding minority youth population, all portend a deleterious gang situation. Barring any major political and economic changes...an increase in gangs and gang crimes seems likely.[1]

Predictions are dangerous, of course, but in this chapter I want to look at several factors that I think will characterize gangs and gang control for the immediate future. They are among the sources of my pessimism: a misplaced emphasis on networking and information sharing among enforcement agencies, the continuing contribution of urban decay to gang formation, and the diffusion of gang culture across the nation.

Information Sharing

Los Angeles Deputy Sheriff Sergeant Wes McBride began systematically collecting gang information in the early 1970s while assigned to an area in East Los Angeles. He developed an operational definition of street gangs that still is largely the core of the definition used by his and other departments (including the LAPD). McBride and his LAPD counterpart, Sergeant Bob Jackson, wrote an influential practical guide to gang identification and control, the only book written by police officers I've ever seen on the topic. In 1977, McBride was instrumental in establishing the Southern California Gang Investigators' Association (later renamed the California Association). He continues to serve as its president after all these years. Finally, it was McBride, along with Jackson, who initiated the cross-jurisdictional gang-rostering system to which I will refer more extensively a bit later.

In sum, this is Wes McBride's legacy: More than any other single police officer in America, Wes McBride has marked, shaped, and driven information sharing about gangs. I admire his work, but I also worry about its ultimate effect, because the models he developed are becoming national models that, to serve narrow enforcement purposes, may inadvertently be spreading street gang culture. To his continuing credit, Wes McBride is more aware of this problem than most of those who enthusiastically follow his lead.

What has happened? Every month, the California Gang Investigators' Association (CGIA) meets to share notes on the latest gang predations and the latest enforcement countermoves. Attendance at the meetings in Southern California alone often exceeds one hundred officers from as many as thirty police, probation, and parole jurisdictions. State and federal officers often come as well.

Special seminars are also held. One of these, in 1989, originally had 387 preregistered participants, requiring a second hotel to handle the overflow. Eventually, almost 500 attendees were registered. A "gang listings" book (price $8.00) was available. Training committees and seminars have been organized. The business of police gang control has been established, along with what amounts to a gang cop profession.

And of course, there are spin-offs. There's the Northern California Association, a chapter in the "Inland Empire" of San Bernardino, Riverside, and Imperial counties surrounding Los Angeles; the Asian Gang Investigators' Association (with 350 attendees at a San

Francisco meeting in 1991); the Midwest Gang Investigators' Association, which recently held its third annual meeting in Minnesota; and currently an attempt to organize a Florida Association. There may well be others of which I'm unaware.

What's the problem? The association members are sharing pertinent information on gangs, gang crimes, current warrants, procedures being used in the fight against gangs, and new legislation. Some investigations are thereby improved. Professional skills and knowledge are increased. Camaraderie is being developed. To read the monthly minutes of the CGIA is to peer through a window on modern gangs and gang enforcement. It looks great, it feels great, and gang cop cohesiveness is considerably enhanced.

But to what end? There are occasional breaks in cases, aids in individual investigations, joint investigative efforts initiated. Whether the hundreds of monthly "man hours" (person hours, if you prefer, although female gang cops are few and far between) are best spent in these meetings cannot be measured, but at least this is open to question. But that's not my concern.

My concern is that all this networking is disseminating, reifying, and distorting the general image of street gangs. The 2,000-member Midwest Association has chapters in Illinois, Indiana, Iowa, Michigan, Minnesota, Ohio, and Wisconsin. The cops' narrow image of gangs is becoming predominant, and it's an image leading to narrow notions of how to respond to street gangs. It's the crime fighter's image that loses almost all interest in why gangs emerge, what they offer to their members, what their neighborhood context is, and what their communities could do about them. For the sake of camaraderie and occasional investigative gains, a constrained image of street gangs is fostered and reinforced. If the newly appointed FBI agent is told to learn about street gangs, where is he going to hang out, at the community agency consortium or the gang investigators' meeting?

But it's not just these offshoots of well-meaning police networking that concern me. More deliberate efforts at training have become institutionalized. Teams of gang experts from Chicago and Los Angeles fly across the country to present gang seminars (remember the LASD captain's training talk—"I'll tell you what a gang is; it's a group of thugs. They're hoodlums, they're crooks and criminals."). The FBI in Washington transmits its conspiratorial, drug-connection view to all area officers and thereby shapes the form and substance of local descriptions reported back to Washington. The DEA and ATF (Bureau of Alcohol, Tobacco and Firearms) disseminate their equally narrow, RICO-relevant characterizations of street gangs and, of course, concern themselves only with enforcement responses.

A National Center for Gang Policy has been established in Washington, D.C., a private clearinghouse for information dissemination with political, justice system, and academic support given to its director, a former Peace Corps worker and community advocate for the old Vice Lords of Chicago.

The Office of Juvenile Justice and Delinquency Prevention (OJJDP) has funded two efforts. The first is a national clearinghouse of gang data and reports, the National Youth Gang Information Center. In addition, OJJDP has funded the Federal Law Enforcement Training Center to provide scores of gangs-and-drugs training seminars across the country. I heard two of the trainers give a presentation on the program and was genuinely shocked to learn the degree to which they equated the gang and drug problems in a professionally slick, convincing format. Only the well-inoculated could come away with any doubts about the intimate control of drug distribution by American street gangs.

Not to be outdone, Wes McBride's CGIA joined with ATF to present a "National Gang Seminar" in March 1992 and another in 1993. Three and a half days of instruction for law enforcement officials (identification required) also provide easy access to Disneyland. The gang business is in full flower.

Somewhere in this are the academics, for they too are spreading their version of the street gang, sometimes succumbing to the audience-pleasing war stories. I've done it—often. So have Spergel, Moore, Huff, Fagan, Skolnick, Hagedorn, Taylor, Vigil, Maxson, Chin, and others. I wish I could know the effect, but my usual pessimism tells me that our message is being lost in our numbers. What remains when the microphone and camcorder are turned off is the fact of gang "spread," not the need for gang knowledge. We, too, contribute to the diffusion of gang culture.

The other form of information sharing is more insidious and raises somewhat different issues. Again, Wes McBride has proved to be a pivotal force. In the mid-1980s, he moved the LAPD and the Sheriff's Department into a proposal to a state agency to put their gang rosters onto a computer system that would permit instant access. Gang rosters, to that point, had consisted of paper files with only local investigative utility. The proposal was approved and led to the Gang Reporting, Evaluation, and Tracking (GREAT) system. The keeper and handler was, and is, Wes McBride.

Over 100,000 gang members are listed in the GREAT system. Because of the reported migration of Los Angeles gang members to other jurisdictions, access to the system has been facilitated for other enforcement agencies. By 1992, there were 150 agencies. The Bureau of Alcohol, Tobacco, and Firearms is hooked in and very involved. Police departments as far away as Louisiana, Texas, Georgia, Oregon, and Alabama are connected. The Honolulu department, to save costs, has loaded the entire GREAT roster onto its own version of the system.

GREAT has about 150 fields of information per gang member, with 38 indexed for immediate access. There is a five-year purge criterion if a member remains clean, but this is five years of "street time" (excluding periods of incarceration). Any agency interested in access to the system can receive training from McBride's colleagues. Included is the issue of entry criteria, those definitional characteristics of suspects that must be satisfied before they can be entered into GREAT. Until very recently, these characteristics were rather ambiguous. With the advent of the STEP act, the Sheriff's Department has tightened the criteria somewhat, based on its own approach:

1. When an individual admits membership in a gang.
2. When a reliable informant identifies an individual as a gang member.
3. When an informant of previously untested reliability identifies an individual as a gang member and it is corroborated by independent information.
4. When an individual resides in or frequents a particular gang's area and affects their style of dress, use of hand signs, symbols, or tattoos and associates with known gang members.
5. When an individual has been arrested several times in the company of identified gang members for offenses which are consistent with usual gang activity.
6. When there are strong indications that an individual has a close relationship with a gang but does not fit the above criteria, he shall be identified as a "gang associate."[2]

As the reader can see, there remains considerable ambiguity in this definitional system, as indeed must be the case. Street gangs themselves seldom have rosters, dues, clear membership classes, and other accoutrements of social clubs or professional associations. Membership is ephemeral; turnover is relatively high; and group-versus-individual behaviors are difficult to classify. For instance, a Florida statewide task force has recommended a GREAT-style rostering system for an estimated ten thousand gang members, which admittedly includes members of "graffiti gangs," that is, taggers, not street gangs.

The ambiguities necessarily involved in rosters like GREAT require careful attention to legal safeguards. My impression from many conversations with enforcement officials is that interjurisdictional gang roster systems, be they GREAT or any variations of it, are embraced as a new weapon with too little attention to their true operational utility. Texas, Nevada, Hawaii, Colorado, Florida, and Illinois, for example, have gang databases similar to GREAT. They are new enforcement toys, and expensive ones at that. Consider the following as typical problems:

- In an earlier chapter, I described how wildly incorrect ATF's depiction of Los Angeles gangs was. ATF is on the GREAT system; how well can we trust that its misidentifications won't affect other agencies' use of its data?

- Asian gangs are sufficiently different from others (and from one
  another, when one considers Vietnamese versus Chinese versus
  Korean, etc.) that system entry criteria are problematic. They
  exhibit different crime patterns, styles of identification (dress,
  tattoos, and so on), place the family name before the given
  name, have multiple spellings for the same name sounds (e.g.,
  Nguyen and Huynh), often have estimated or false dates of
  birth if immigrants, and so on. The potential for identification
  errors is not trivial.
- Membership turnover will lead to an inflation of numbers and
  the misidentification of former members as still current.
- Is gang membership a relevant legal status in all participating
  states, or is there a serious civil liberties issue here?
- Will the identification of a gang migrant from one city increase
  police surveillance/harassment unfairly in his or her new city?
  Similarly, does entry on an interjurisdictional roster unfairly
  increase prosecutability merely because of gang membership?
- Will defense attorneys have equal access to roster information
  (inquiries, identification, etc.) on their clients and associates?

A letter from a colleague in prosecution circles considered what
had been learned in that state about gangs and migration and
appended this note to a review of police agencies' use of such data:
"Finally—you'll like this—most didn't often have to contact other
agencies about gangs, but ALL thought that regional, statewide, and
national databases would be VERY USEFUL. I can only admit that you
told me they wanted toys!!" In California, the Department of Justice is
working toward a combined California–Nevada network that will permit access to a quarter of a million purported gang members. Some
toy!

People in the prediction business identify two problem areas,
labeled *false positives* and *false negatives*. In gang identification, false
positives are persons entered in the system as street gang members
who really are not, like friends of members, former members, members of nongang groups such as taggers, or local peer groups who copy
some gang symbols or occasionally commit delinquent acts. False negatives are bona fide members unknown as such to the police, like
younger or recent recruits, those who have avoided arrest or successfully misidentified themselves, and others able so far not to fit the
entry criteria of the system involved. The GREAT system has been
criticized for allowing errors of both types, but the number of errors is
itself almost impossible to calculate. The critics are equally vulnerable
to criticism. The task is to find ways to minimize errors and protect
individual liberties as well as to facilitate legitimate investigation and
prosecution.

The task is far from easy. It is unclear that the financial and personnel costs will not outweigh the advantages to enforcement. But

the gang roster movement is strong and has now reached the federal level in the form of coordinated planning, legislation introduced in both houses of Congress, and the initiation of congressional subcommittee considerations. A 139-page report on the 1992 subcommittee hearings is available from the U.S. Government Printing Office. The subcommittee assistant counsel responded with a June 1993 memo expressing grave concern about the FBI proposal to establish a national gang roster. Senator Dennis de Concini of Arizona introduced a bill to establish a national gang roster system. The model considered, of course, is GREAT.

The Government Accounting Office (GAO) reviewed the GREAT system, finding it generally acceptable, although in need of safeguards in recording agencies' entry into the system and use of its information. It is ATF that is planning this national system, a fact that does little to decrease my anxiety. The cost was initially estimated at $8 million, more if mug shots are added to the system, as proposed. Senator de Concini asserted that "Los Angeles cannot continue to provide nationwide service without the assistance of the federal government. While the problem of gang violence has historically been regarded as a local problem, it must now be recognized as a national menace."[3] This article citing the senator's concern goes on to make it clear that the rationale for the national system is based on the facts and myths of the Los Angeles gang–drug infiltration of the nation. The same old roundup of the usual suspects is invoked—Bloods and Crips in Phoenix, Portland, Seattle, and Shreveport.

We now stand, it seems, on the threshold of a national gang roster system, based on GREAT, without anything approaching an evaluation of its utility to enforcement or its endangerment of civil liberties. Even a careful assessment of the number and importance of system errors—false positives and negatives—has yet to be undertaken. Yet an interagency plan is well under way, involving the ATF, FBI, National Law Enforcement Telecommunication System (NLETS), and National Criminal Identification Center (NCIC). NCIC Regional Working Groups (RWGs) and an interagency Gang File Working Group (GFWG) are creating the Gang Information Network (GIN). When all these acronyms come together, they're still talking about the GREAT "toy," with the Sheriff's Department's criteria of membership and the STEP act's definition of a gang ("three or more people") still in place. The aim remains suppression only.

The Urban Underclass and Gang Proliferation

The future of the American street gang—in particular, predicting its increase or decrease—will depend on a number of factors. Public policy trends is certainly one of these but is difficult to anticipate beyond a few years. The place of firearms is another, since guns yield violence, more guns yield more violence, and violence is one defining aspect of our thinking about gangs. New control techniques is another possibility; one could attempt different predictions based on the future ascendance of street work, suppression, and community organization approaches.

I want to concentrate on two matters that seem to me to have become transcendent. The second of these, the diffusion of culture, I will discuss a bit later. The first, the effect of the increasing urban underclass, remains in my mind the foremost cause of the recent proliferation of gangs and the likely best predictor of its continuation.

I must admit I came to the connection of gang proliferation with the urban underclass rather slowly. As the number of emerging gang cities came to my attention over the years, I tried to account for this growth in a rather unsystematic way. I was aware that the inner cities of America were becoming less inhabitable. Social services were receiving less and less federal and state support. Unemployment among minority youth clearly was increasing. Many inner-city families that could afford to move out were doing so. An inner-city vacuum was developing at the same time that more cities were emerging as gang cities. I saw the parallel and started to talk about it, stressing the issue of youth unemployment and racial or ethnic segregation, the latter because white gangs were disappearing in a plethora of black and Hispanic gangs.

I wasn't being terribly systematic in all this, however, until the publication of sociologist William Julius Wilson's book, *The Truly Disadvantaged*.[4] Whatever may be the criticisms of Wilson's thesis about the urban underclass, his argument remains compelling. Although it is derived from his work based in Chicago, it generalizes to many other urban settings. What I'll do here is present major propositions in Wilson's portrayal and then describe a conceptual model of the way that these propositions in turn may relate to the growth of gangs.

The essence of the urban underclass is often stated as the emergence in the inner city of "persistent and pervasive poverty." The basic propositions include the following of interest to us:

1. The industrial base of urban centers has steadily been removed. Big plants and small have closed down, transferring their operations to suburban centers, industrial parks, and, in some instances, other countries. Thus many of the jobs that provided more or less steady employment for ghetto families have disappeared. Behind them is left an overbalance of service jobs, from fast-food cashiers to social worker positions, for which many inner-city residents, especially younger males, are ill prepared.

2. With the relative success of the civil rights movement in weakening the barriers to jobs, property ownership, and educational access, there has been an out-migration of middle-class blacks and Hispanics from the inner city to suburban areas. They have taken with them their middle-class institutions—active churches, children's agencies, business clubs—and their experience in community organization. Left behind is a higher proportion of lower- and working-class families in a declining job situation, with fewer practical skills in community organization.
3. Despite civil rights gains, many urban areas have experienced a relative increase in minority segregation, producing ever more distinct pockets of concentrated minority populations.
4. Public school systems, mirroring the falling resources for health and welfare, have increasingly failed to respond to and prepare inner-city youths for life in a setting of service shifts and high school credentials for the entry-level positions still available.
5. The effects of job loss, class and race segregation, and school failure have put extra strains on both family life and the availability of suitable male partners for would-be wives and mothers. More homes empty of parents or surrogates during the day leave more youths on their own, without the structure of school or work to occupy their time and hone their skills. Poverty becomes more prevalent and more persistent.

Obviously, Wilson's elaborations on this picture are worth the reader's attention, and other references fill the picture out.[5] My purpose, here, however, is to describe how the growing underclass problem is connected with the gang situation. Part of the answer, quite obviously, is simply a response to poverty.

One of my favorite black gang members put it simply: "Bein' poor's a mother-fucker." Black or Hispanic, the hurt is sorely felt. Joan Moore, writing from the perspective of East Los Angeles's Hispanic areas, comments:

> If a community's economy is not based solidly on wages and salaries, other economies will begin to develop. Welfare, bartering, informal economic arrangements, and illegal economies become substitutes—simply because people must find a way to live. Young people growing up in these communities have little to look forward to.[6]

Both Moore and Jeffrey Fagan point to the gang–drug sales correlation and suggest that the relationship is not causal but spurious; the growth in both is a function of a third underlying variable, namely, the growth in poverty occasioned by the acceleration of the urban underclass.[7]

The Webster Commission, investigating the 1992 "spring unrest" (riot, rebellion, or civil disturbance, depending on your viewpoint) following the first Rodney King beating verdicts referred back to the 1965 Watts riots and noted balefully that federal investments had, since 1981, again declined: from $23 billion to $8 billion for job training; from $6 billion to zero for general revenue sharing; and from $21 billion to $14 billion for local economic development. Finally, "overall federal support for housing programs has been cut by 80 percent. As a result, the great gulf between rich and poor has continued to widen."[8] The commission was not aware of but would have found congenial an analysis by Thomas J. Bernard of how the "truly disadvantaged" of Wilson's model are propelled toward what Bernard calls "angry aggression."[9]

My young gang member was crude but on the mark in feeling the "mother-fucker" quality of poverty in the ghetto. The gang member's situation has shown no improvement in the period between the two Los Angeles riots. What I said after the first is just as true now:

> Our gang member is thrown into his group. He is frustrated, insecure, and trapped in his environment. He is not having much fun, although he makes much of the enjoyments he finds. He is a rationalizer and a self-deceiver in his attempt to get through his adolescence with as few psychological scars as possible. He is not the man he would like to be, and his search for peer status is seldom adequately rewarded. Sad to say, society's most effective mechanism for transforming him—sheer maturity—is not under its control.[10]

Poverty is powerful. But it's not poverty alone that's involved in the growth of gangs. Even the poverty responsible for the growth of urban crime is not alone responsible for the growth of gangs. We get close with the work of John Hagedorn. While I was fumbling with my notions of unemployment and segregation, Hagedorn applied Wilson's underclass formulation to the appearance of street gangs in the 1980s in Milwaukee, a "rust-belt" city that mirrored the changes that Wilson noted.[11]

Hagedorn stresses the "deindustrialization" common to many cities and can be clearly documented in Milwaukee. This is Wilson's notion of the shift from manufacturing to service jobs. He also notes the increasing segregation in his city and the out-migration of middle-class residents which leaves the area with a residue of less skilled, less empowered residents. All this, plus the increase in poverty, creates "a generation on the way down."

Undereducated, underemployed young males turn to the illegal economies enhanced by gang membership, including selling drugs in some instances. Older males who in earlier decades would have "matured" into more steady jobs and family roles hang on to the gang structure by default. The newer gang cities like Milwaukee thus

emerge, looking much like the traditional gang cities. Hagedorn concludes:

> Wilson's view...that the underclass was made up of people who are
> "collectively different" from poor people in the past, is based on his
> research in Chicago, where the underclass, and its gangs, have had sever-
> al decades to become entrenched. That process, we suggest, is beginning
> in Milwaukee and in other smaller cities. Gangs today in small cities are a
> red flag telling us the underclass is not just a problem for the New Yorks
> and Chicagos.[12]

I noted in an earlier chapter the expansion of the age bracket of
street gang membership, into the twenties and thirties. Adults and
kids alike are affected by the closing off of nongang opportunities.
Wilson put it succinctly following the 1992 Los Angeles riots:

> The most fundamental change is that many poor black neighborhoods
> today are no longer organized around work. A majority of adults in inner-
> city ghettos are either unemployed or have dropped out of the labor force.
> Consequently their everyday lives are divorced from the rhythm and real-
> ity of the American mainstream.[13]

And for those coming up, Wilson notes, "In his time, Martin Luther
King could still have a dream. These kids can't even imagine a
future."[14]

I want to take the reader through a slightly complex chain of reasoning in order to make the connection between the underclass and
gang growth different from an analogous connection between the
underclass and crime or violence generally. I've made the point several times that street gangs are something special, something qualitatively different from other groups and from other categories of
lawbreakers. Yes, I believe there is a causal connection between
underclass and gang growth, but it can best be understood by asking
first what is needed in any city, at a minimum, to permit the formation of street gangs.[15]

One problem here, since we define street gangs partly in terms of
their illegal behavior, is that we might actually be predicting
increased crime levels, not the emergence of gangs specifically.[16] A
different approach, therefore, is one that starts by specifying the characteristics of gangs, so that we might then ask what underclass variables should best predict the emergence of gangs in some cities but
not others. This is the approach taken here.

Characteristics of Gangs

I will set out two broad categories of variables that may be seen as
minima for gangs and also point out that two classes of variables need
to be distinguished, those leading to the (1) onset and (2) mainte-

nance of gang structures. The first category is basically structural and the second basically psychological. Both are derived from a combination of personal observations of street gangs over many years in many settings and the professional literature on gangs.

Structural Factors

American street gangs consist of minorities, the young, and usually males and are found principally in the working- and lower-class sections of their cities. Structural underclass variables used to explain and predict the emergence of gangs should relate to these characteristics. That is, a relevant question is, What is it about the growth of the urban underclass that might exacerbate the coming together of lower-

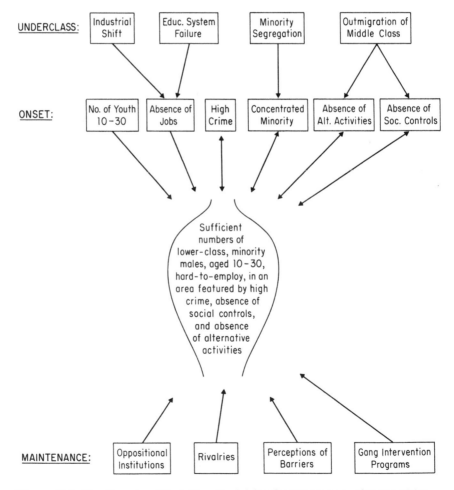

Figure 7-1. The Structural Variables Model for the Emergence of Gang Cities.

class, male, minority youth in urban settings that would result in the formation of street gangs?

First, I'll list the pertinent structural variables (one shouldn't take the term *structural* too literally) and then suggest how they might be patterned in Figure 7-1.

Urban Underclass Variables (from Wilson)

1. Out-migration of minority middle class.
2. Shift in industrial modes from manufacturing to service.
3. Segregation of minorities.
4. Failure of the educational system.

Proximal Gang Onset Variables

1. Sufficient number of minority youth, that is, ten to thirty.
2. Absence of appropriate jobs.
3. Absence of acceptable alternative activities.
4. Concentrated minority populations.
5. Comparatively high crime rate.
6. Absence of community and informal controls.

Proximal Maintenance Variables

1. Oppositional structures and institutions (police, school officials, etc.).
2. Rival groups (gangs, in most cases).
3. Shared perceptions of barriers to improvement.
4. Gang intervention programs.

Figure 7-1 is a first attempt to illustrate how these variables may promote (onset) and reinforce (maintenance) gang structures. The gang structure is the same as that in Chapter 3, Figure 3-1. Note that there are reciprocal relationships between the gang and several of these variables. That is, the gang can be reinforced by having its own effect on not only the maintenance variables but also selected onset variables, as indicated by the double arrows in Figure 7-1.

Several comments about Figure 7-1 are in order. The underclass variables are drawn as causing the onset variables, but one could argue about the directness of the causal paths and about other under-class variables. The reciprocal relations listed (double-headed arrows) are only the most obvious. Also, maintenance variables are probably the most difficult to measure without adopting ethnographic proce-dures. So for now, we're really predicting here to onset and, I hope, some time-ordered reciprocality.

So far, then, I've suggested a set of structural and community variables that, should they become more severe over time, might pre-dict the onset of street gangs in cities with a burgeoning urban under-class problem. We can now add to this a very different category of variables, those describing the psychological characteristics of gang members that seem to separate them quantitatively from nongang or (in some instances) youth peripherally involved in gangs. Following this depiction, yet another task faces us, that of asking what structural variables might affect youth positions on these variables. For example, if gang youths have comparatively less impulse control, why should their inner-city setting cause an even greater loss of impulse control or larger numbers of youths with low impulse control?

These final variables are seen primarily as onset variables, in the sense that they lead local youth to "try on" gang life and find it psychologically satisfying. Many local youths hang around gang members for a while and then stop. My concern is with those for whom there is a good fit between individual needs and gang (not just group) attributes.

Psychological Factors

Street gang members were once characterized (e.g., Frederic Thrasher) a bit like happy-go-lucky urchins. Modern gang researchers do not report this impression. Also, they were at times (e.g., Yablonsky) described as violent and sociopathic, following megalomaniacal leaders. Modern gang researchers have eschewed this picture as well. Rather, a consensus has slowly been achieved that although gang members enjoy access to some satisfying social connections and share commitments to a set of group or subcultural values, they nevertheless lead comparatively dissatisfying lives that prepare them poorly for mainstream adult life.

Although most gang research has been carried out by sociologists, individual traits and characteristics have attracted some attention— enough to suggest modal gang member positions on a number of dimensions. From the depictions in Chapter 3, I'll list only the more prominent of these, not as hypotheses or predictions, but as research conclusions. Remember, however, that there are major variances around the modes.

Status Needs

The most consistent characteristic derived from major gang writings is the member's need for social status, a need fulfilled in large part by affiliation with a street gang. Belonging to a well-known, tough, "bad" group provides immediate and repeated gratification.

Identity and Belonging Needs

In much the same way, the gang provides an otherwise absent sense of "who I am" by offering an identity with a group, claims to reputation, and a neighborhood known for its place as gang territory. It's not "I am Carlos" but "I am Maravilla" or "I am a Crip" that takes over the gang member's self-esteem. For a youngster unsure of himself, this is powerful stuff.

Social Disabilities

Work done in Chicago highlighted a set of social disabilities that featured gang members considerably more than their nongang counterparts. Such disabilities range from poor table manners to inadequate conversational skills with adults. Such disabilities reduce one's capacity to adapt to school, maintain prosocial adult liaisons, find and hold jobs, and interact effectively with the police. The implicit consequence is that youngsters with lower social skills find greater acceptance among themselves and shy away from the sorts of contacts that might reduce their gang commitments. Some confirmation comes from Gerard's description of core gang members as (1) feeling that "others are hostile or indifferent to his welfare" and as (2) having "marked inferiority feelings with regard to conventional activities outside the home."[17]

Bases for Affiliative Needs

In addition to comparing gang with nongang members, or simply describing gang members noncomparatively, we can infer some of the differences between gang and nongang youths by comparing core with fringe members on the assumption that distinguishing features of these can be extrapolated to the gang–nongang differential. Recall from Chapter 3 that core and fringe members do not differ on various sociodemographic variables. But a factor analysis of a wide variety of psychological and behavioral characteristics yielded two independent factors of interest:

The *"deficient-aggressive" factor* showed core members to have greater deficits in measured intelligence, impulse control, school performance, group dependence, and desire for change while at the same time being more committed to or involved in aggressive and criminal behavior than fringe members were. Core membership, it seems, attracts the less socially capable and more antisocial youngsters.

The *"group involvement" factor* referred to group-qua-group behaviors. Core members' participation in both formal and informal gang activities showed greater involvement in cliques and also greater commitment to the group than fringe members displayed.

I summarized these factors by saying that "the gang is seen as an aggregate of individuals held together more by their own shared incapacities than by mutual goals. Primarily, group identification is important as it serves individual needs; it leads to delinquent group activity only secondarily and only in the absence of prosocial alternatives."[18]

I don't claim any clear independence of these characterizations of status and identity needs, social disabilities, and group affiliation factors. In fact, they seem quite complementary. But they are of a different order from the structural factors that Wilson offered. The conceptual task that these more psychological factors present to us is explaining how they might become exacerbated in the growing underclass setting. In group terms, what is it about this urban context that leads gang members to make of their companions a normative reference group to the virtual exclusion of all others?

Referring back to Figure 7-1 to ask which variables might most directly affect these psychological characteristics, I would nominate the educational failure, minority segregation, middle-class out-migration, and another factor not found in the list—parental competence (broadly conceived). These more or less directly affect youth characteristics, rather than being mediated through the onset variables noted in Figure 7-1. Thus a full causal model written in broad categories would look something like Figure 7-2.

Summary

The scheme presented in these pages asks first what it is about gangs that seems required for their formation and then what it is in the urban underclass matrix that would strengthen these gang-promoting factors. It is unclear whether static levels of the underclass are sufficient to predict the proliferation of gangs. My bet, certainly, would be on the side of dynamic measures; that is, the growth in the urban underclass is more predictive of gang onset than its absolute level is. In addition, from the potential gang recruit's viewpoint, it is unclear

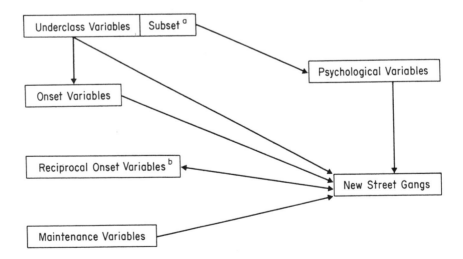

Figure 7-2. Broad Category Causal Model.

[a]The subset of underclass variables are those directly affecting psychological factors.
[b]The reciprocal onset variables are a subset of the full list.

whether the absolute level of deprivation or the relative level of deprivation is more important. A static or absolute answer to the underclass question might favor the absolute level for the individual; the growth prediction for the underclass question might favor relative deprivation for the individual.

In any case, what I've presented is a working model for understanding the astounding proliferation of American street gangs. It starts by asking what it takes to foster gangs—the words inside the gang structure in Figure 7-1—and then whether urban underclass variables can be used to produce these formative factors. Finally, the model adds the importance of gang–underclass reciprocity and also the importance of gang maintenance factors. One of these, the police response, we extensively described in Chapter 6's discussion of enforcement suppression programs.

I do want to emphasize one other point about Figure 7-1 and the fact that these psychological variables are not included in it. The point is that different facets of street gang formation—who joins, how gangs get going in a community, and how they are maintained once they have formed—are not necessarily determined by the same factors. Covey and his colleagues came to the same realization, noting:

> The best explanations of how and why juvenile gangs form may have little to do with why individuals join gangs, although one cannot occur without the other. Once formed, juvenile gangs may persist or disintegrate for reasons that may have little to do with why they formed in the first place.[19]

Support

Does the model work? Does it successfully predict or explain the proliferation of street gangs? I can't be sure. It feels right to me; it makes sense in terms of everything I've come to understand about street gangs and the settings in which they've become established. But the model needs a lot of empirical testing.

Still, there is some support for the overall idea. Hagedorn's work in Milwaukee, though only one case study, nevertheless is compelling as one piece of evidence. Ronald Huff's work in Ohio provides some validation. A static underclass situation in Columbus is associated with some gang emergence, whereas a dynamic, worsening underclass in Cleveland is associated with the emergence of a larger number of gangs. Much as described in Milwaukee, "normal" youth groups—break-dancing groups, local street corner groups—developed rivalries that eventually turned into street gangs. Although there was some gang leader migration from Chicago and Los Angeles, Huff found no evidence that the local gangs were chapters of the bigger-city gangs. Rather, Columbus and Cleveland, like most other cities, have produced their own gangs.[20]

A more direct test was provided by Pamela Irving Jackson, who found two variables predictive of gang cities, job opportunities and proportion of youth population (ages 15 to 24).[21] She noted the support that this offers to Hagedorn's connection between underclass and gang. She also found that crime levels and gang presence were not predicted in the same way, which provides some support for my contention that gang proliferation is not merely a mirror image of other urban ills, that to some extent it has causes different from those of crime per se.

Another test is beginning to take shape in work I've begun with Jeffrey Fagan at Rutgers University. The test is limited for now to the one hundred largest U.S. cities and attempts to predict the time of gang onset. (Since most of these cities are now gang cities, predicting gang versus nongang is, sadly, no longer possible.) As I write this, only the very first analysis has been undertaken, and Fagan is preparing our first public report.

Using 1970 census and labor statistics, racial segregation significantly predicts post-1980 gang emergence, as does a low proportion of persons in the labor force. Moreover, an interaction between loss of manufacturing and unemployment also predicts the emergence of gangs. Concentration of poverty does not make a difference.

Two points are of interest in this initial pattern, which otherwise generally supports the underclass hypothesis. First, it is the static rather than the change measures that are more powerful. I'm not sure at this point what to make of this. Second, the findings hold up for blacks but not for Hispanics. Wilson's work, of course, was based on blacks. Hagedorn's and Huff's descriptions are primarily of black, not Hispanic, gangs. Data from the Maxson–Klein gang migration study suggest that the recent proliferation of gangs has disproportionately taken place in cities with a preponderance of black rather than Hispanic gangs.

Thus it may turn out, although we're grasping at straws with the current data, that the underclass hypothesis will be more pertinent to black gang situations. In her introduction to John Hagedorn's *People and Folks*, Joan Moore suggests that the two groups indicate different routes to the urban underclass. In the West, where Hispanic gangs are more numerous, many of the new cities are boomtowns rather than deindustrialized rust-belt towns. But the booms didn't reach the inner cities, and in the 1980s Mexican immigration overstocked the local labor markets. The poorest and least prepared were closed out of legal economies. This leaves "leftover" young Hispanics comparable, for gang purposes, to the leftover blacks of Wilson's Chicago, Hagedorn's Milwaukee, and Huff's Ohio cities.

A recent paper by Curry and Spergel also suggests this ethnic differential: underclass theory for blacks and subculture theory for Hispanics and institutional racism and pervasive alienation for blacks and expanding cultural gang traditions for Hispanics, the result of what Vigil refers to as the multiple marginality of Hispanic youth in the Southwest. Still, even though there are somewhat different ethnic routes to the development of gangs, the outcome is nevertheless similar.[22]

There may be a number of cities for which the underclass hypothesis is inadequate. The underclass variables may well be relevant but less powerful. Hispanic gang cities might be included, and smaller cities and towns might be included. The next section of this chapter offers a second hypothesis that helps fill the gap. But so far, it seems worthwhile to note that at least some portion of the gang proliferation problem is reflective of larger social ills. Merely addressing gang problems through gang intervention, be it street work or suppression, won't have much effect. In a rare application of his thinking specifically to the gang issue, Wilson's view is summarized: "Youth gangs can best be addressed by addressing the larger problems in a more comprehensive way, thus helping the ghetto underclass and significantly reducing youth gangs, whose membership is comprised largely of the inner city minority underclass."[23]

The Diffusion of Street Gang Culture

If the urban underclass does not account equally for black and Hispanic gang proliferation—and that is merely a suggestion, not a finding—it also seems a bit difficult to apply it with equal relevance to larger, chronic gang cities and smaller, newer gang cities. Recall that the bulk of newer gang cities have emerged since 1980 and even more so after 1984. With the dividing line at populations of 100,000, 56 percent of larger cities with street gangs have emerged since 1984, and 76 percent of the smaller cities with street gangs emerged after 1984. In addition, the emerging gang cities have produced, on the average, fewer gangs and smaller gangs in smaller inner-city areas.

This does not invalidate the relevance of the underclass hypotheses to smaller and newer gang cities. It does, however, make one look about for additional generic contributing factors.

When I started my national interviews with gang cops, I had no special interest in the notion of cultural diffusion. But I soon learned its importance. And when I started writing this book, I had no intention of writing about taggers other than to specify their exclusion from my street gang characterization, along with skinheads, stoners, bikers, and so on. But again, I've had to revise my thinking. The present intrudes rapidly. The cultural diffusion of gang styles has had, I now believe, a major impact on gang proliferation. Taggers have emerged as a highly visible example of that process. We can learn something about the process by looking at the transformation of some taggers (and I emphasize *some*) from individual adventurers to mem-

bers of rival gangs. Once again, the concept of group or gang cohesiveness is critical.

I mean nothing obscure and very sophisticated by the term *cultural diffusion*. It is similar to the notion of *social mimicry* used in ethology,
the borrowing by one species of animals of the social behavior of
another that has survival or another advantage.[24] In the earlier section on information sharing, I described the way that gang experts
have been disseminating information on gang patterns and behavior
styles.

We can add to this the diffusion by way of the media. Television
specials on gang behavior are now common, as are depictions of gang
members in TV dramas. News programs are replete with gang incidents, often showing youngsters in the background in their special
clothing, flashing gang signs, and framed by the camera against a
graffiti-filled wall.

Cable channels and clothing commercials add to the dissemination by emphasizing inner-city symbols, styles of dancing and strut-
ting, and suggestions of gang dress (sports caps and jackets from the
Los Angeles Raiders, baggy and sagging pants, and so on). MTV, in
particular, has become a major carrier of youth culture.

Movies like *Colors*, *New Jack City*, and *Boyz 'n the Hood* depict
romanticized versions of gang life, with all the stylistic accoutrements
they can add. In a phone conversation I had with a gang cop in Texas,
he blamed *Boyz 'n the Hood* for perpetuating gang culture in his community. When I pointed out that this particular film had a particularly
strong antigang message—just about all the gangbangers are sad cases
who end up killed or maimed—the officer countered by saying his
gang members concentrated their attention on just one scene, ignoring the overall message. In that scene, a gangbanging subhero follows
up the automatic weapons shooting of several rivals by slowly and
coolly approaching each on the ground and applying a single head
shot to each. He then calmly returns to his car and drives off. "That's
what the gang guys get off on," said the officer, "That was a righteous
gang execution, and they really grab their jocks on that one; forget
the rest of the movie."

His comment struck me forcibly. A strong antigang message was
again contorted to fit in with gang culture, just as the reaction to an
Operation Hammer arrest could be turned into a contribution to gang
cohesiveness. We need to keep in mind how gang members and
potential members become the subjects of their own gang dynamics.
When the popular rap artist Ice T writes a descriptive *Playboy* article
on the Los Angeles gang scene, he contributes to these dynamics even
while asserting his own graduation from gangbanging and concern for
gang peace. Writing in the gang vernacular spreads the culture.

Youth culture is a powerful set of symbols and language.
Remember "valley-speak?" The San Fernando Valley father said to his

son, "Two words I don't want to hear from you—one is `gross' and the other is `awesome.'" "For sure, Dad," responded the youngster. "What are they?"

If there's an appealing stylistic pattern, it will be picked up and incorporated. As late as the early 1980s, Anne Campbell and Steven Mincer drafted a paper on the dissemination of youth and gang sub-cultures in which they suggested the absence of gang culture diffusion.

> Youth in one part of the country are relatively ignorant of others' activities until it reaches the point of mass movement or violence. The net effect is that New York teenagers, already factioned within the city into their own areas, have virtually no knowledge of the situation of gang members in Chicago, Los Angeles, or Philadelphia.[25]

If that statement were true in the early 1980s, then how dramatically things have changed! In May 1993, there was even a highly publicized "gang summit" held in Kansas City, attended by purported current and former gang leaders from coast to coast. I keep thinking of my invitation to visit "nongang" Pittsburgh, where I found so many school kids playing with gang styles—the names, the reported initiations. They were wannabes without any local gangs to mimic, yet they had the essentials thanks to MTV, the media, and all the rest.

As I noted earlier, the danger here is something like behavioral contagion: If you walk like a duck, talk like a duck, and act like a duck, you may end up a duck. Milwaukee's gang founders started in break-dancing groups that represented their neighborhoods. After entering contests, some of these developed rivalries attached to their areas, and these in turn evolved into classical street gangs with the traditional gang's age-graded group structure. Then they became, in name, associated with the Chicago gang nations, People and Folks. You don't have to have a real "nation" affiliation; you just need to adopt the cultural signs. Notes David Dawley, a 1960s gang organizer with the Vice Lords:

> Cultural symbols and associations run through our culture very fast through music and television. There are Crips in Portland that can't find Los Angeles on a street map and Vice Lords in Columbus or Cleveland who have never been to Chicago, and wouldn't let a Chicago Vice Lord come in and tell them what to do. But they share the symbols and the legends.[26]

Such a culture need not be limited to inner-city areas. Keep in mind that as cities expand, as busing and school magnet programs break down neighborhood enclaves, youth and families from the ghetto reach the suburbs. Culture travels by media and also by personal carriers. "T.L. is a middle-class youth who sports a black Los

Angeles Raiders cap and cutoff pants and punctuates his speech with
gang-style street slang," starts a *Los Angeles Times* article entitled
"Hard-Core Gangs Attract Middle-Class Imitators" (December 25,
1989).

California's Office of Criminal Justice Planning publishes a
newsletter called *Newsline*. Its summer 1990 issue stresses the emergence of suburban ganglike groups: "In Lancaster, about 50 miles
north of Los Angeles, 200 students threw stones at a policeman who
had been called to help enforce a ban of the gang outfits that have
become a fad on some campuses."[27] It is only one example among
many of the "copycat" groups and social mimicry. There are few
youths in this country by now who can't describe something about
gang attire, symbols, names, and the like. They can walk and talk and
act like ducks. Rivalries develop to solidify group structures. Because
school and police officials react to copycat groups as if they were
street gangs, they inadvertently help complete the transformation.[28]
And nowhere can that process be seen more clearly, or more omi-
nously, than with taggers.

From Chaka to Tagbangers

A few years ago, a young man from East Los Angeles left his home
every day, spray-paint supply at the ready, to find walls, bridges,
signs, and vehicles on which to spray his "tag"—CHAKA. It seemed to
be his only goal, his compulsion, an addiction to tagging. His work
was not only local. Chaka tags appeared up and down the West Coast,
and soon Chaka took on the trappings of a folk hero, the lone rebel
foiling the combined police forces of California.

He was the ultimate in finding individual expression. He tagged,
but he didn't fight, he didn't steal, he didn't harass. Chaka was the
all-American, one-man counterculture, James Dean with a spray can.
Eventually, he was apprehended, brought to court, charged with six-
figure damage, and convicted. On his way out of court, into the elevator, down the hall, and out of the court building, his legend seemed
about to die. But after he had left, the elevator revealed that legends
never die; there was CHAKA, painted inside the elevator door. Or so
goes the story, anyway. So striking was this young man's impact on
the public imagination that the period since has been characterized in
a newspaper report as the "post-Chaka era."

Chaka was one of many thousands of youth across the country
who have become taggers. The staid *Smithsonian* magazine devoted a
full article to them in July 1993. For the first few years, the rebellious
but individual nature of tagging was constant. Tags became common
nuisances to the public, a street art form with its own "can control"
technology, a persistent bother to property owners whose walls were
targeted, but no threat to their lives.

Creator, a 13-year old tagger, received a full write-up in the *Los Angeles Times*. But so did a disgruntled businessman whose walls were violated four times in one day, each time following a cleaning of the walls. Creator and the businessman, individuals in conflict.

Like Milwaukee's break-dancers, taggers began to form groups, called *tagger crews* or *tagger posses*. Creator reportedly joined three crews. In New York, in September 1990, a tagger crew "who pursued dancing and spray-painting graffiti as hobbies," was indicted for a mugging murder in a Manhattan subway station. New York subways are perhaps the most tagged location in the nation. This murder could just have been an aberration, a one-time escalation of a simple altercation. Any group, in the right set of circumstances, can become violent. This crew, the FTS crew, had heretofore been unknown to the New York police.

But as the reader may anticipate, I'm going to describe a growing and disturbing pattern in which tagging turns to ganging. Individual taggers, sharing areas of residence or school attendance, come together in their mutual interest, sharing the adventure of tagging. They become crews or posses. Then in some cases—not most, but some— they take on the trappings of street gangs. They walk and talk and act like ducks. Some become ducks.

From my home in West Los Angeles, I drive to my university along graffiti-lined Jefferson Boulevard. Then, just a few weeks before I wrote this, some form of community action took place, and my trip on a Monday morning took me by miles of painted-over walls. Homes, stores, and school fences all had fresh coats of paint, and the graffiti were nowhere to be seen.

The next morning, one large tag had appeared on a newly painted wall, a three-letter tag clearly not of one of the street gangs. The significant feature, however, was that this tag had been crossed out, and next to it a different three-letter tag had been painted. This is not typical tagger action. Taggers leave one another's tags alone. But gangs regularly express territorial rights and challenges by crossing out and replacing rival gang names. Here, we had two taggers (or tagger crews—only they can know) showing the strictly ganglike acts of graffiti warfare.

This is not expected tagger behavior. Taggers are supposed to be into the hip-hop culture, rapping and break-dancing themselves through a brief youth culture into young adulthood. An unexplained killing of a young college-bound, high school student was found to be the result of a tagger crew rivalry. The town was Reseda, in suburban San Fernando Valley, and more than one thousand mourners attended the funeral. No gang wake, this, but a middle-class gathering in search of an answer.

The answer is that tagging is coming to the edge of gangbanging; *tagbanging* has become the compromise term. In Long Beach, police report thirty tagger crews causing $500,000 in cleanup costs per year.[29] The crews are developing territories, some overlapping with gang territories. Some crossover membership is reported by police, although most taggers remain "graffiti vandals," in one cop's words.

To learn more about tagging, the Los Angeles Sheriff's Department mounted a "sting" operation in Long Beach, posing as a British film crew doing a piece on taggers. Advertisements brought in dozens of taggers to tell their stories on camera, yielding sixty-five videotaped interviews. The LASD's intelligence files now have the stories, the faces, and the names. The most boastful of the taggers soon saw the inside of a courtroom. In Orange County, two undercover police officers posed as schoolboy taggers in two Anaheim high schools. By participating in some tagging episodes, they gathered the intelligence needed for a tagger sweep that netted seventeen suspects.

Tags are usually drawn with ordinary spray paint. In the face of local ordinances making spray cans less available, substitutes are easily found. Oven and glass cleaner bottles can be used up close; crayons are allowed. Liquid shoe polish can be substituted for paint. The law interferes, but it does not prevent. A challenge is a challenge; the spirit of Chaka lives on.

Most tagger crews remain true to their cause: "We just write, man. Gangs are into killing," one youngster reported. They know the difference. Another had passed the tipping point and knew it: "We're not a tagger crew no more—we're a gang." The use of firearms seems to be the dividing line between taggers and tagbangers. It's very much like my insistence that a distinguishing feature of street gangs is their own recognition of their delinquent or criminal commitment. Crews become gangs, if they do, as a result of violence between rivals. Sullivan and Silverstein, the former a sheriff's deputy who alerted me to the aforementioned sting operation, recently estimated thirty thousand taggers in Los Angeles County, with several murders attributed to some who have become more ganglike in manners and weapons.[30]

In many city areas, including gang areas, tags rather than gang graffiti have become predominant. The reader can spot the difference readily in most cases. Look for three-letter tags; KWS (Kings with Style), TIK (Think I Care?), EWF (Every Woman's Fantasy), and OFA (Out Fucking Around) are typical. These are not territorial names; there are no Crip attachments or Vice Lord attachments.[31] The names are statements of position, of rebellion, of autonomy, of youth cultural style: NSC (No Self Control), WNA (We're No Angels), YGW (Youth Gone Wild), and LOC (Lunatics on Crack). They're having fun with their names, saying something about themselves rather than claiming turf: KMT (Kings Making Trouble), ATC (Addicted to Crime), BAC (Beyond All Control), and YPN (Your Property Next).

Tagger crews are not gangs in my terms. They don't meet the criteria, spelled out in earlier chapters. But some of them have become ganglike, and others have been transformed into street gangs. The shift from the one culture to the other seems not to be a difficult one. The KMT crew recently was the recipient of a drive-by shooting. Seven were hit, and the payback cycle was off and running. The TIKs have firearms available but are reluctant to use them. Still, if they continue to taunt the USCs (Unstoppable Criminals), there's every reason to expect their guns to come out and another series of paybacks to take place. In the city of Bell, KWS took aim against such traditional street gangs as Maravilla and Florencia. The first KWS fatality soon followed. After the arrest of two M2K (Mob to Kill) taggers for murder, Long Beach police reclassified M2K from a tagger crew to a criminal gang.

This taggers-to-tagbangers-to-gangs progression presents not only a clear example of youth cultures at work and not only a confirmation of the importance of group cohesiveness to the formation of gangs; it also presents a pervasive and difficult challenge to communities and public authorities. As usual, it's not the communities that are responding but the authorities alone.

The California Gang Investigators' Association, Wes McBride's large collection of gang cops, and others in the business have now formed TAGNET, the Tagger and Graffiti Network Enforcement Team. At the members' meeting in January 1992, reports spoke of a melange of tagging, burglary of stores for spray cans, stabbings, shootings, deaths, and crew/gang connections. California cops recognize the vandal nature of most taggers, but they're also adopting a frame of mind that could enhance reactions to taggers as gang members.

A more recent monthly meeting of TAGNET yielded a picture of the growing bridging of tagger crews and street gangs. TAGNET members reported:

- "DTS (Destroy to Survive) began as a tagging crew and is now acting as a traditional street gang. Bangers are beginning to join with taggers because tagging is fun.'"
- One city reports the existence of twenty-five tagger crews, another more than seventy.
- A major police department claims a "correlation" between gang members and taggers in a particular housing project area.
- Another jurisdiction reports the arrest of four taggers with guns, and a fighting rivalry between two tagger crews.
- A small but active gang city reports that its taggers have joined together to fight a crew in another city.
- Thirty-four crews have been identified in a suburban city not known for gang activity.
- A middle-class suburban community reports gang members and taggers running together, with some taggers initiated into a Hispanic gang.

- A sheriff's station reports, "Twelve of the Hispanic gangs and seven black gangs have affiliations with tagger crews. Two crews are doing shootings, not graffiti or tagging."

Meanwhile, the legislative machine is plugged in and warming up. Twenty graffiti-related bills were introduced in the California state legislature during the 1992 session. Local ordinances in various cities are increasing the penalties for graffiti writing, offering rewards for information on taggers, and requiring stores to lock up their spray cans, outlawing sales to minors. Maybe we'll see special sentence enhancements for murders committed by taggers! The tagger group NFR (No Fucking Respect) was recently targeted with the STEP act process.

Even if the whole reservoir of potential gangs were exhausted in the next years, other group formations like taggers provide a new potential supply for the proliferation of gangs. The prospects are not good. And if our communities and officials respond to such groups as they have to gangs, with direct attacks in the absence of concerns for prevention, then the prospects will worsen. I don't want to be a purveyor of gloom, but it does seem to me to be getting darker out there. I see no signs that gangbanging will peak soon.

8

Is the American Street Gang Unique?

Although the Skins make good copy ... the real threat is the gangs,
[which] have effective organization and money, as well as strength of pur-
pose to protect their own families and property.

> Marie Douglas, "Auslander Raus! Nazi Raus! An Observation
> of German Skins and Jugendgangen"

Representative Joseph Kennedy, the former president's nephew, visited Berlin in late 1992 following a series of vicious skinhead attacks on
ethnic minorities in Germany. Echoing his uncle's famous "Ich bin ein
Berliner" speech, the younger Kennedy presented his own warning to
the German government that had been so slow to respond to these
skinhead attacks: "Ich bin ein auslander (foreigner)." Appearing soon
after on CNN's Larry King show, Kennedy took a phone call from a
listener who asked whether Kennedy didn't think we ought to be
working on the control of our own gangs here in the United States
first?

The inquiry raised again the question of whether ganging in one
form or another is a relatively universal phenomenon or whether the
American gang is special to our country. Can we learn from the experiences of other nations? Can they learn from us? On some topics,
after all, we Americans have been better at telling other nations how
to get their house in order than we are at handling our own social
problems.

How might we answer the question of whether or not American
street gangs are unique? There are structures called gangs in many,
many countries, ranging from the two vicious brothers called a noto-

rious gang in China, to the relatively benign youth network in a Swedish town, to the organized criminal *yakuza* of Japan. The issue might best be resolved by looking for typical characteristics of American street gangs that can be found in these structures in other countries.

- Are they territorial, with gang rivalries?
- Do they acknowledge themselves as criminally oriented groups?
- Do they exhibit cafeteria-style crime patterns rather than crime-specific focuses?
- Are their structures moderately cohesive, neither very tight and structured nor quite amorphous?
- Are they the seeming products of current or growing inner-city areas of poverty, alienation, and discrimination; do they display the cultural signs that provide special identification—the attire, tattoos, argot, and overall posture of the street gang?

Most writers, in describing gangs elsewhere and in other times, have not raised the issue in this fashion. In most but not all cases, we cannot be very sure that the gangs referred to would fit our pattern. Recall from Chapter 3 that Geis traced gang terminology back to Chaucer in 1390 and Shakespeare in about 1600 and cited gangs earlier in our own century in the United States, Germany, France, England, and the Soviet Union.

The *barras* in Argentina contained fewer than eight boys each and usually fewer. They are described as relatively leaderless but tightly cohesive, with a special gang argot. Rivalries and cafeteria-style crime patterns were common. These *barras* were street gangs, though not clearly in the American style.[1]

A description of the French situation in the 1960s was rather deliberately written in such a way as to invalidate a direct French–U.S. comparison, stressing subculture more than group structure. Nonetheless, the description of one group does contain many of the U.S. elements of the traditional gang of considerable size, with subgroups and temporal durations.[2]

T. R. Fyvel, describing the British teddy boy subculture in the late 1950s, saw strong parallels with other gang reports—very sketchy reports—from such disparate locations as London, Liverpool, Daytona Beach, New York, Duisburg (Germany), Vienna, Paris, and Stockholm. The descriptions really do not merit any decision about the structural characteristics of the groups. Rather, the more common bond seems merely to be their boisterous behavior in scattered incidents.[3]

Without the slightest attention to the generic characteristics of gangs and with considerable comingling of such terms as *gangs, peer groups, delinquent youth groups, youth groups, youth gangs, criminal youth gangs,* and so on, Clinard and Abbott twenty years ago compiled an extensive list of affected locations. These included the United States, England, France, Germany, Sweden, Japan, South Africa, Zambia, Dakar, Kinshasa, the Cameroons, Ceylon, Thailand, Malaysia, Chile, Argentina, India, and Egypt. The sense of a worldwide phenomenon is shaky at best, however, given no structural depiction or other defining characteristics.[4]

More recently, but again without the data in many cases to confirm the presence or absence of American-style forms, Spergel listed most of Europe, Russia, Kenya, Tanzania, South Africa, Australia, Mexico, Brazil, Peru, Taiwan, South Korea, Hong Kong, China, and Japan. Another list was provided by Covey, Menard, and Franzese in a very useful overview of the cross-national situation. Here we find Ghana, Montreal, Australia, Puerto Rico, Jamaica, the Netherlands, Italy, Yugoslavia, India, Indonesia, and Thailand, among others.[5] The Ghanian description, in particular, strikes me as very much in the American street gang style.

In the process of preparing this book, I've come across a number of gang descriptions provided by colleagues. These will help zero in a bit more on the supposed uniqueness of our own street gangs. I'll also describe the situation in several locations I've visited in order to test out this uniqueness in a more structured fashion.

Belgium

A Belgian university colleague described Brussels as currently having five street gangs that exhibit intergang rivalries and cafeteria-style crime patterns. Membership consists principally of minority youth— second- and third-generation Turks and Moroccans—living close to the center of the city and committing their crimes nearby. This description of American-style gangs should be kept in mind, as it resembles closely those for several cities to be listed later. The Belgian criminologist offering this depiction also reported several smaller specialist youth gangs concentrating mainly on thefts from automobiles.

Spain

Although youth gangs of some sort have been described in Spain in the past, few seem to exist now. "Bandas juveniles" articles in the 1980s reported variously on right-wing extremist groups ("rock-a-billy gangs"), a few gangs in which females seemed to provide the principal source of cohesiveness, and "la banda del Paco" (Paco's gang), a single group of four young teenagers. In Madrid, youths from marginalized neighborhoods on the edge of the city met in correctional institutions and afterwards formed some short-lived gang structures that caused troubles in the city center. Similar reports came from Valencia. The area of the Atocha train station in Madrid recently reported one group consistently involved in minor thefts and vandalism, along with harassment of train and bus station travelers, sometimes with a show of knives. Otherwise, both Madrid and Barcelona now seem to be free of gang activity, however defined, except for hooligans associated with sports teams.

Slovenia

"Gangs" of juveniles coming before the urban court are described as short-lived, small, and loosely structured, more youth networks than gangs (see the upcoming parallel with Sarnecki's groups in Sweden).

Canada

Montreal has been described as having Haitian gangs, but on being challenged for particulars, the reports more closely fit small cliques of delinquent associates. Vancouver and more recently Toronto have offered reports of Asian gangs that are similar to such reports in the United States (Chinese in Vancouver, Vietnamese in Toronto).

France

The *blousons noirs* of earlier years disappeared. Now new American copycats called *casseurs* have formed in Paris and Marseilles. Angry youngsters, mostly Algerian or other African blacks, are forcing a discussion of gangs versus youth culture. A translation of a French magazine article offers the following:

> Today the gangs define themselves by territory and ethnicity. We are approaching what is happening in America. In the Paris region, the gangs of young blacks have multiplied. They are called "zoulous." Their general quarters are the Halles, La Defense, and Trocadero. They have their code, language and uniform. They wear baseball caps, too large jeans, bomber-jackets, Adidas shoes. They listen to rap music. They make themselves noticed in the subway. They consider themselves warriors of Bronx. They dream of a mythical America. There are about 100 different gangs in the region of Paris. There is aggression in buses and in the subway. One group of arrested gang members admitted having taken part in over 200 acts of aggression.[6]

Mexico

Because the Cholo culture of our own southwestern Hispanic gangs derives from Mexico, that country might well be expected to contain similar structures. Mexico City, it appears, more than fills the bill, with a variety of small and large street gangs spawned by the abject poverty of one of the world's most populous and most poorly serviced cities. Street life there makes ours look inviting. Territoriality, gang rivalries, graffiti writing, versatility in delinquent activities, character-istic clothing, and other gang culture symbols all combine to suggest a close analogue to many U.S. gangs.

Japan

The famous *yakuza*, organized criminal gangs, have received all the attention in Japan, but they are not street gangs in any sense. If any street gangs exist, they have not managed to gain attention in the context of these traditional crime structures. There was one brief arti-cle, however, in the *International Herald Tribune* (October 6, 1989) that caught my eye:

> The police in Osaka, Japan, have charged 13 gangsters with moon-lighting as car thieves in addition to their regular gang activities. Police quoted them as saying they had to steal overtime in order to make ends meet, the *Japan Economic Journal* said. The newspaper reported one gang-ster as saying, "We couldn't put food on the table working as regular hoods."

South Africa

Here old black gangs, where they existed, were uprooted by the Group Areas Act, apartheid's program for relocating blacks into new, segregated townships and autonomous areas. New gangs quickly formed in these new areas, variously based on earlier ties, prison con-nections, residential proximity, and the disorganized controls of the new settlements. The illegal economy is often the only available means of support; cohesiveness and rivalries flourish. I've been unable to flush out any research on these situations, other than that of local news writers. The research challenge would be considerable.

The Philippines

Twenty years ago I heard a description of Manila gangs that included more violence than I had ever imagined. Little has changed, except that old turf wars have settled down and there is now a wide variety of street gangs, adult criminal gangs, specialty cliques, and criminal syndicates comparable to the *yakuza*. The gangs are said to be ubiqui-tous, their crime patterns highly varied among gang types. But these descriptions are mostly anecdotal, and courageous researchers are needed to separate violent myth from fact.[7]

Hong Kong

In the late 1960s, a detached worker program was organized to respond to local neighborhood groups that seemed to be approaching ganglike proportions. Territoriality, rivalries, and subgrouping all were noted. Criminal behaviors were, for the most part, relatively minor, and social and sports activities were engaged in quite commonly. The criminal orientation "tipping point" seemed not to have been reached at the time of these reports.

China

When I asked about gangs on my first visit to China, officials would respond only with the case of two brothers who had become criminally notorious. In those days, the denial of a disapproved phenomenon had little meaning; it could be that it didn't exist, that there was no formal knowledge of it, or that it would not be admitted publicly. In 1989, however, a Chinese professor visiting my university agreed to share lunch and "secrets." He described small specialty gangs (theft, drugs) and larger structures with versatile (cafeteria-style) crime patterns. Some were territorial, he said, but others were not. Smaller ones tended to be restricted by age, and larger ones were more heterogeneous regarding age. He noted that at the time, his descriptions were the only ones of their kind, as no one else in China was doing any research on gangs.

A start has now been made by two scholars undertaking a longitudinal study of teenaged gang leaders.[8] This is significant at least in the public acknowledgment of such groups, but the available report says absolutely nothing about the gangs themselves.

New Zealand

Brief reports cite the involvement of Maori (native aboriginal) gangs in serious violence in recent years. A Maori legislator visited Los Angeles to learn from the experts there how New Zealand might best respond. One hopes he also heard of the failures in Los Angeles.

Australia

Kenneth Polk, a longtime colleague who left the University of Oregon to take a position at the University of Melbourne, told me about a ganglike situation there that closely resembled that in Stockholm at the opposite point on the globe, which I will describe later in this chapter. He provided an extended description for use in this book, and I cite excerpts from it because it refers so directly to material covered here earlier and also to more about to be covered:

> Here these groups are likely to flow into available public spaces of various kinds, and these provide the locales for the conflicts which occur between groups. Alcohol seems to play a role in the general pattern of

recreation, and then becomes a visible feature of violent confrontations, including those which result in a killing.

The common elements between the U.S. and Australia, however, are important. The violence is definitively masculine, involves persons of underclass or working class origins. . . .

A major source of economic pressure experienced by underclass young persons in both the U.S. and Australia is the significant structural change in the labour force that has resulted in the rapid loss of entry-level jobs for those with little to offer in the way of qualifications, skills or experience. Historically, in the days of maximum industrialization, large numbers of adolescents were able to exit school during or immediately after secondary schooling, and find a wide range of work in manufacturing and even in various parts of the service sector. Loss of these jobs in the post-industrial phase is catastrophic not because it means that there is a temporary period of unemployment, but because the work involved constituted the entry portals into adult work careers. For large numbers, the resultant unemployment is exceptionally long-term in its duration.

The result is that this group of young people suffers from a significant disruption of personal and social development. Unable to move from school life to early work careers, they find closed to them the sequences which follow on from entry into work. Work provides a personal identity and access to social groupings crucial for such following steps as meeting potential mates for family formation. Access to the wage is essential for any process of legitimate independence and movement into the surrounding social world as a developing adult.

Those experiencing the resultant disruption in social transition suffer, then, a number of problems. They are, as it were, stuck in a kind of social and psychological no-man's-land. They lack the economic resources which might ease and legitimate the movement away from the parental household (and, accordingly, it is not uncommon for considerable strain within the family to develop as the presence of a non-productive emerging adult continues to exert pressure on family resources). How does one spend his or her time during the day, if one is not at work or not in school? Where is the young person to find social associations, security, any sense of fun and excitement?

Dumped out into the streets, it appears that Australian underclass young people, rather than staying close to their home "turf, " are more likely to seek out places to spend their time that make minimal demands on their scarce economic resources.

It is our perception that they pool together with others like themselves, and flow into the available public space, often collecting up with others as they move. Thus, their experiences are collective youth experiences, but the groups are informal and brittle in terms of their social composition, often breaking down and being added to as the group moves through space. A common feature of their movement is the use of public transport, including trains, buses and trams. Their destinations are likely to be such hubs as train stations, shopping malls, parks and reserves, the central and local shopping centers, and if they have a bit of money, one or another of the pubs along the way that they may come to identify as their "local."

Rather than being based in a localized "turf," Australian underclass youth are likely to move routinely through public spaces. They seek low or no cost entertainment, excitement and fun. A definitive feature of their movement is that its limits are established by their lack of resources.

The various movements in the public space can be viewed as providing a series of points of frictional contact. At one level, for example, there is literally the competition for space with other users of public spaces. In shopping locations, for example, shopkeepers are likely to become agitated when the underclass youth are seen to compete with potential shoppers in such a way that such shoppers might become fearful, and seek other shopping venues. Police are frequently called in at that point, and expected somehow to free up the space by moving the young people along....

There are particular features of the frictional movement which carry a high potential for trouble and violence. When the mix of conventional citizens drops off, as in the late hours, and when the young people have been drinking for some time, then the target for one group's violence is most likely to be other marginal or underclass individuals or groups. Insults are quickly exchanged between groups, and a contest of honour is staged. A fight may follow, most commonly involving fists, but occasionally knives. Either way, the results in some cases will be lethal.

The forms of collective violence in Australia, then, appear to have a somewhat different cast than found in the U.S. The conflicts are not neighbourhood based, nor in any sense constitute a defence of "turf." Instead, they can be seen most often as arising from frictional contacts arising out of the social movement of young males through public spaces (and thus they are likely to have a highly episodic appearance).

When serious violence results, it tends to be a "flare up" which gets out of hand, as opposed to systematic and sustained violence between two defined groups. In the few instances where lethal violence results from something like a feud which extends over time, the groups are still informally and loosely organized, and would see themselves as tied in simple friendship as opposed to being united by a bond of common, formalized gang membership.

No gangs, here—not in our sense. But note several features of these youth groupings: They move from their own neighborhoods into others to commit their offenses. Rivalries have not developed, and violence is episodic and reactive to the borrowed "space" of the moment. Important to this account is Polk's comment on the informally organized but persistent fashion in which these underclass youths leave their areas in groups, by public transportation to public commercial areas, and then return home by public transport following the evening's depredations. It is a style of action that we will meet again in Stockholm, Zurich, Frankfurt, and Stuttgart.

Papua New Guinea

I admit to chuckling when it was first suggested that traditional gangs were to be found in Port Moresby, but I soon apologized in the face of several independent reports—one from a former student of mine who took a leave of absence there, another from a local police officer, and others from research by the Australian Institute of Criminology.

Originally of tribal origin but increasingly urbanized in a city of more than 120,000 inhabitants, the gangs are named groups —KKK, Mafia, Six Male Tigers, Texas, and others—that were first noticed in 1963. By the mid-1980s, estimates placed the number of street gangs at around fifty, these ranging in size from ten to one hundred members, with one having as many as five hundred.

Often referred to as *rascal gangs* because they were originally more boisterous than criminal, Port Moresby's gangs have developed traditional structures very much like the traditional American structure. These groups tend to be territorial, with rivalries. The core–fringe depiction is apt, with subgroups, some initiation rites, and moderate levels of cohesiveness. Criminal versatility includes theft, fencing, breaking and entering, vandalism, extortion, auto theft, and the drug trade. Violence is not a common feature of gang activity, but overall levels of crime lead to more than twice the arrest rate of nongang arrested youth. By the 1980s, these groups are said to have become more consolidated, structured, and deliberately criminal in the absence of effective police and community controls. All in all, the pattern here joins those in Brussels and Mexico City as good analogues to the U.S. pattern.

Russia

As a member of a juvenile justice delegation to the Soviet Union a few years ago, I obviously kept my ears open for any mention of gangs. It was in the early and exciting years of Gorbachev's *glasnost*, and our Soviet hosts were being surprisingly open about their difficulties. An official from the National Procuracy (prosecutors' office) commented, "We borrow the worst from the West," so I had hopes. Soon we heard familiar words: *mafia* (a term the Soviets use broadly for any organized crime pattern), *punkers, rockers, hippies*. Then came a comment about territorial gangs in the city of Kazan in the Volga region. There were, it was said, many gang fights: "almost civil war in that city." Suppressive police action, in the absence of community controls, had merely driven the gangs underground.

Further inquiry suggested that gangs had proliferated in a number of cities: Kazan, Mozshansk, Alapaevsk, Shumerlya, Neftekamsk, Taganrog, Ivanovo, Dzerzhinsk, Kaluga, and Volgograd (formerly Stalingrad). Causes were listed as "social differentiation and deprivation" (in our terms, ethnic segregation and poverty, by now a familiar story). As in many American cases, social groups evolved into criminally oriented street gangs, with ethnically based territoriality, notable group process and leadership, and intergang rivalries. These are not, it was emphasized, merely delinquent groups.

According to a high-ranking police official in St. Petersburg, the MVD (state Interior Ministry police) now contains gang specialists as a consequence of the gang activity. As in the United States, gang cops lead a frustrating life in the street. There was a report of two hundred persons in a gang fight, of whom only one or two were adjudicated. Many assault complaints were simply not followed up, some with questionable reasons: "Someone lost an eye, and . . . was . . . informed that he had apparently run into a tree" (*Izvestiya*, February 9, 1989).

During my visit, I was approached by Elena Savinkova, a staff member of the Procuracy Research Institute. She had been studying the gang situation mainly on her own without colleagues with whom to share her work. We have corresponded ever since, and much of my picture of Russian (no longer Soviet) gangs comes from her.

There is a wide variety of *informal youth groups* (IYG), as they are called. In describing the proliferation of gangs in many Volga cities and others, some overstatement has been made if one keeps in mind my restricted meaning of street gangs. In addition to nontroublesome groups, Russian investigators have labeled various manifestations of Western youth culture types: sports fanatics, bikers, punks, heavy metal groups, and neo-Nazis. There also are elitist youth groups who lord it over less successful peers.

Street gangs as I have described them, however, have indeed evolved in Volga cities with minority populations such as Tatars. Kazan, Tomsk, and Naberejnie Chelni each have yielded research descriptions of "the Kazan phenomenon," that is, street gangs as we know them. They are territorial with strong intergang rivalries and are large structures with age-graded subgroups containing age-specific leaders. Cohesiveness seems quite strong, more so than in many U.S. gangs perhaps, strong enough to indoctrinate younger and fringe ("husk") members in the group's mores and to exact membership dues.

In Kazan, a city of 1.5 million, gang membership has grown over ten years to an estimated 20,000 or more, a number similar in ratio to what one might find in Chicago or Los Angeles. By the time they reach this level, IYGs have been involved in a wide variety of crimes, including considerable violence—"the dominant character of their activity is criminal."

Savinkova notes causal factors that are virtually synonymous with those attributed to the formation of American gangs. She also suggests that these groups have proved highly resistant to intervention efforts and faults her Russian welfare and police colleagues for launching programs without an adequate understanding of group process in gangs. Obviously, she echoes much that I have argued in this book. We can add these Russian cities to Brussels, Mexico City, Port Moresby, and perhaps Manila and South Africa as locations of American-style street gangs.

Sweden

We come now to my most deliberate attempt to assess the uniqueness (or not) of our American gangs. During my 1985 sabbatical leave in Stockholm, it became clear that this "Venice of the North" was free of any street gang (*gatugang*) activity, although some skinheads were episodically active.

The awkward note in this description was research by Jerzy Sarnecki on gang behavior in an anonymous town north of Stockholm. On further investigation, however, it seemed that the term *gang* was being used very loosely indeed. What Sarnecki's very good work referred to was loose networks of youths, unstable in membership and relatively minor in crime involvement, that did nonetheless yield data on companionship patterns in delinquent incidents. These were not gangs. The town turned out to be Borlange, a small rural city in an area (Dalarna) known best for its picturesque scenery and its distinctive arts and crafts. Even if it were transported to the United States, Borlange would be hard put to come up with a street gang.

Sarnecki, colleague and friend, has since dropped the gang terminology:[9] The way things are going, he may get to use it again. During a visit several years later, Stockholm had developed seriously delinquent groups, serious enough to have led to a volunteer street work program ("Mothers and Fathers on the Town") and a clique structure documentation system in the police department's juvenile division.[10]

Residentially located in a housing development in Rinkeby on the outskirts of Stockholm, the groups were composed of a mixture of second-generation Turks, Yugoslavs, Moroccans, and others, the children of imported "guest workers" and asylum seekers common to much of western Europe. On weekends, much like the Australian youths, scores of Rinkeby youths poured into the subway leading in short time to the center of Stockholm—the central station and underground area called Sergelstorg and Gamla Stan (Stockholm's Old Town).[11] There, especially on Friday and Saturday nights, they gathered in small groups, occasionally preying on local residents and tourists alike in muggings and harassment. Smash-and-grab break-ins of central city stores were common, along with general carousing in the downtown area, daytime robberies and shoplifting, and rolling of drunks. This is less a "cafeteria style" than U.S. gangs exhibit. And unlike U.S. gangs, after an evening's spree, their Swedish counterparts took the subway back home.

I went down into the lower level of Sergelstorg with the clique-documenting juvenile officer, but there was no need to point out these youths; they were obvious and, late at night, were not there for the simpler joys of life. Their way of eyeing passersby was clearly one of looking for their next mark.

A colleague supplied the subtitle for an article that my wife and I wrote on this phenomenon in several cities: "Commuting to Turf." It's an apt phrase. These immigrant youths are strongly group oriented but nonterritorial. Home is home, not territory. Their area of group activity is several miles away, the central city.

Even though they are described as having a special, mixed-language argot and secret signs, these groups have no rivalries and little hierarchical or age-graded structure. A key explanation lies in the residential patterns of Stockholm and other European cities compared with their U.S. counterparts. In many European cities, housing developments are placed in the outskirts rather than the inner city. Urban underclasses are residentially located, by design, in the suburbs. These immigrant minorities, in many respects analogous to our own inner-city Hispanics and blacks, are geographically less well situated to form rival street gangs. Rinkeby itself is a mixed bag of open food markets, ethnic stores, modern if stark apartment buildings, open spaces, and lots of shoppers from miles around. Name plates of individual apartments show Swedish, Finnish, Turkish, Moroccan, Greek, Yugoslavian, and Ethiopian names, among others. Such mixed ethnicity does not easily produce street gangs, which traditionally are quite ethnically homogeneous.

In data from the Stockholm Project, a large longitudinal study of youths in the city, Rinkeby stands out as one of two areas with the highest levels of delinquency and claims of gang membership, but the latter is low, and few can provide a gang name.[12]

By 1992 and my most recent visit, more changes had taken place. Rinkeby is less of a single source; other areas are now the targets, in addition to downtown, where the police presence has been increased, including a substation on the lower level of Sergelstorg. But the very latest news is the appearance of "Svenska tagger crews" in 1993. I took pictures of local tags late in 1992—now there are crews.

Frankfurt

The significance of the "commuting-to-turf" pattern of Melbourne and Stockholm is that it is not unique. Several other cities were nominated as potential gang sites by colleagues, and I visited them late in 1992. We were given the Frankfurt tour by an undercover narcotics and juvenile officer and met with others as well. Some 20 percent of Frankfurt's population is minority—Turk and Moroccan guest workers, producing second-generation delinquent youth groups almost identical to the Stockholm pattern. There is the same suburban housing, the same subway rides to downtown Frankfurt on weekends, the same crime patterns, and the same returns to the housing developments in Griesheim, Nied, and Nieder-Eschbach. These groups used to have names, but they learned from the response by the police to drop their names, thereby making it more difficult to distinguish *jugen-cliquen* from *jugenbanden* (cliques from gangs). The principal crimes listed are robberies/muggings, thefts, some assaults, and drug sales (hashish and heroin).[13] There is no territoriality, and there are no gang rivalries.

Stuttgart

The pattern here, again, was the same; I won't bother to repeat it. The additional facet was the event-based appearance of soccer hooligans. Stuttgart has a perennially successful team that acts as a magnet for fanatics from all over the state of Baden-Wurtemburg. However, the more stable problem is the commuting delinquents, serious enough to have led, as in Stockholm, to a street worker program and a youth violence unit in the police department.

Zurich

I have a written report on youth gangs in the glorious old tourist town of Lucerne in Switzerland, but it turned out to refer to cliques of polit-ically active street youths. I expected to find nothing more serious in Zurich, a town of banks and stores catering to Europe's wealthiest customers. But *jungenbanden* exist even there, similar to those in the German cities. The home countries differ somewhat—Turkey, Yugoslavia, Albania, Lebanon, and Chile—with first- as well as sec-ond-generation youth comprising the groups and residences both twenty minutes outside town but also in town. The crime patterns are essentially the same, and as elsewhere, there is no territoriality or intergang rivalry.

Words like *homeboys* and *hip-hoppers* reflect the cultural infusion, as do the Raider caps and other paraphernalia in all of these cities. In Zurich, the police talk about these groups as "toys" or toy gangsters, imitating the U.S. gang culture seen on TV. Their crime patterns are worrisome, but not the core of group life. They are some distance, as yet, from the tipping point of American-style ganging.

London

My colleague David Farrington at Cambridge University has been a faithful dispenser of cross-Atlantic reports on gangs in London and Manchester. I did not find the American street gang but, rather, a well-formed pattern of drug gangs, known locally as "Yardies." English ganglike structures in the past—teddy boys, mods and rock-ers, skinheads—have by and large been closer to youth culture move-ments than to discrete peer groups. Today, as well, the "gang" pattern differs from ours, but also from those of the England of yesteryear.

In London, there are a few Chinese organized-crime groups tied to the Hong Kong triads. A few white copycat groups have appeared and disappeared, and a few Indian and Pakistani groups have carried their back-home disputes onto the British Isles. But the main action is with the Yardies. These are Jamaican ("Afro-Caribbean" in current English parlance) drug distribution and sales gangs. They are described as relatively unstable in structure and leadership, with only a few rivalries based on sales territories. Crack has only recently come to London, and in small amounts; hash and heroin are the main drugs. There is some evidence of the Stockholm pattern, but many dealers are in the central city as well.

Everyone with whom I spoke—Scotland Yard, the National Criminal Intelligence Service, a criminologist, and a street ethnographer—was in agreement. London's problem, and not a major one, is the drug gang. Street gangs in our sense are not to be found. A few areas, such as Brixton and the Mozart estate ("Crack City") in West London, have had off again–on again flirtations with crack sellers, but these have not been street gang operations as we know them. Yet as recently as December 1993, suspected Jamaican Yardies were being intercepted at the airport and returned home in a new and highly controversial crackdown. The National Criminal Intelligence Service has reported ten crack-related murders in London for the year 1993.

Manchester

This city is a sprawling, red-brick metropolis with little to please the eye. Built more like American cities, it is quite ghettoized. A dense, drug-infested housing area in ill repair, Moss Side, has garnered a great deal of negative publicity as a center for drug dealing and violence. The former award-winning housing project of Hulme is now being torn down as a social planning failure. Alexander Park Estates[14] seems hardly better, and West Gorton is quickly becoming known as the new Moss Side as the old Moss Side residents move in its direction. Rusholme, the area of Pakistani and Indian concentration, is known out of the side of the English mouth as "Khyber Pass" and "Curry Alley." To the visitor, it is a vibrant ethnic community (though reportedly with a taste for hashish).

The gang situation is better defined here than in London. Surrounded by Moss Side police brass and then toured through the worst areas by the divisional drug chief, I learned about the Gooch Close gang and the Cheatham Hill and Pepper Hill gangs. They are strictly drug related, with violence—lots of violence—resulting from arguments over sales territories. There are no known nondrug gang formations beyond a few specialty cliques dealing in car parts. There is, however, a distinctive drug gang dress style: very baggy pants, a leather jacket over a sports-name T-shirt, unlaced sport shoes, and a cap. These guys were easy to spot. Manchester is the closest European city I've seen to the many American cities with their oppressive underclass problems, yet its gangs are organized strictly around drug sales; they are not street gangs in the U.S. style.

But one should not assume that the drug gang problem is any less disturbing than a street gang problem would be. As in Detroit and Washington, ubiquitous drug gangs can yield high levels of violence. An editorial in the *London Mail* on Sunday (October 24, 1993) reported, "Parts of Manchester and London have been likened to (American) Wild West towns." The paper called for a major police crackdown, either a U.S.-style suppression program or a program similar to the draconian Prevention of Terrorism Act enacted in England.

Berlin

This was the last stop on my street gang journey and the city most commonly touted as having American street gangs. In the mid-1980s, right-wing extremist attacks on immigrant youths, most notably Kurds and Turks, led to a counterdevelopment of immigrant youth groups for self-defense purposes. It was just a short distance from this to territorially based *jugendbanden* or *jugendgangen*, youth gangs with well-defined turfs, gang names, and symbols.

As we listened to descriptions from police, researchers, a detached worker agency called Gangway, and a refugee welfare agency, we got the sense of 1960s-style American gangs—traditional structures developing, with rivalries and all the rest, but with the lower violence level of that earlier era than is currently found in the United States. These are youth gangs with as yet little adult participation, with up to fifty active members. There are age-graded subgroups ("babies" up to "killers," we were told). I sketched a traditional structure for the youth violence officers in the police department. With a startled look on their faces, they exclaimed, "Yes, yes, that's it!"

Turkish (often Kurdish) second- and third-generation youths predominate, but the groups also incorporate Yugoslavs, Lebanese, Palestinians, Poles, and some Germans. Along with autonomous or spontaneous groups that rise and fall, there are a few well-established traditional structures—the 36ers (logo in English wall-graffiti), Black Panthers, Fighters, Bulldogs. A recent publication adds Raps, Bomber Boys, Club 7, La Mina, and Aderpower.[15] Some female members are included, and the crime patterns are versatile, including drug use but not sales.

These are not the commuting gangs of Frankfurt, Stuttgart, and Zurich, as they reside well within the sprawling lower-class areas of West Berlin (with some copycatting in the East). Kreuzberg, a lower-class immigrant area of the former West Berlin that seems more suited to the East, is the most active gang area. Gangway is located there, as are the 36ers. The graffiti are remarkably American in style, and the housing stock is typical of many slum areas well known to Americans. Other locations are Wedding and Tiergarten.

Community controls are low, as Germans tend not to get actively involved in social problems in the streets. We were given many sad examples of the Kitty Genovese phenomenon, spectators watching assaults on the weak but failing to intervene. In these circumstances, control is left to an undermanned police force whose three-person youth-violence unit only gathers intelligence and formulates policy. Gangway deliberately divorces itself from gang violence, preferring to work the softer, prevention pole of the intervention continuum. The result, understandably, is that there is much room for intergang and gang-versus-skinhead flare-ups to take shape and add a further measure of cohesiveness to the gangs already formed. There's every reason to expect some of the shorter-lived spontaneous groups to solidify as well. The Bulldogs and Fighters have evolved in this way. Another group, the Barbarians, is said to be on its way, a loose confederation numbering up to two hundred involved youth.

Berlin, then, joins our list of Brussels, Mexico City, Port Moresby, and others. In almost all cases, the background includes ethnic growth and segregation in an impoverished setting. If these latter contexts form as they have in the United States in the undeniable presence of the diffusion of American gang culture, then there is every reason to expect other cities to join the list. The copycats, wannabes, and "toy gangsters" will, via both choice and pressure, escalate into the American pattern. Once in, as we've learned in the United States, most of the forces serve to solidify the situation. It's hard, though not unknown, for a city to retreat from gang to nongang status.

Although the United States can still be judged to contain the great majority of gang-involved cities, other spots have developed their own variants on ganglike structures, and a few, as noted, have given birth to genuine street gangs of the American sort. Our street gang, still numerically predominant, is beginning to look a bit less like ours alone. This is not a development to be applauded. My Russian host said it well, indeed: "We borrow the worst from the West."

Epilogue

Goddammit, shut up! You invited me here to get my view of this situation, and you can do me the courtesy of listening to me—uninterrupted. Then you'll have some facts to use to ask some informed questions for a change.

In the midst of the 1963 controversy over the Group Guidance detached worker program, suspended while the program and the police worked out a compromise regarding its means and goals, a series of seminars was held for some forty juvenile officers. First they heard from the head of the LAPD's Juvenile Division, who lambasted the program to the full and vocal approval of the audience. Then they heard from the director of the program, who received a stony silence at first and then a very vocal rejection of his whole message about working with gangs.

Finally, I was invited to present the researcher's viewpoint and findings to date about the program. I thought of myself as Mr. Neutral, the disinterested bystander offering an objective appraisal, but this audience would not permit that stance. It seems, in the words of Secretary of State John Foster Dulles about nonaligned nations during the cold war, that "it is immoral to be neutral." I could hardly get started before the antagonistic questions and interruptions threatened to destroy my presentation—how could I, they wanted to know, even think of being "objective" about these murderous gang members?

The outburst quoted at the top of the page was mine. And it worked miracles. Forty cops sat up, fell silent, listened attentively, and then asked useful questions—each one preceded by "Sir, what do you think...." That was my first public utterance on gang issues and, with few exceptions, the last for almost the next thirty years. Then in 1992, I found myself "coming out" as a policy-concerned researcher. The hyperbole about gangs and crack, my increasing awareness of gang proliferation, and increasing frustrations with misguided gang suppression combined to send me into a series of not very subtle public statements, especially aimed at some of our government policy-makers. These statements were duly picked up by the media, and the die was cast. This book and the invitation to write it are the culminations of my coming out. Writing has given me the opportunity of expressing my views without, this time, first having to shout, "Goddammit, shut up!"

The federal Administration on Children, Youth and Families (ACYF) is now devoting many millions of dollars to a variety of what the agency calls gang prevention programs. Most of the funding comes from, and is much affected by, the antidrug dollars authorized by Congress. More millions have recently been given to the Office of Juvenile Justice and Delinquency Prevention (OJJDP) to mount new gang prevention and intervention efforts. Will ACYF and OJJDP listen to the sorts of issues raised in this manuscript?

Woven through these pages are four principal themes: gang definitions, gang proliferation, gang cohesiveness, and the drug connection. I'd like to finish our journey by reprising these very briefly.

The definitional issue, which at first might have seemed "only academic" to some readers, turns out to be both ambiguous and critical. It is critical to a clearer understanding of what we are and are not talking about. It allows us, for instance, to draw distinctions between street gangs and drug gangs and thus more clearly understand gang structure, gang migration, and the limited role of crack in gang migration. It also suggests that street gangs and drug gangs may well call for distinctive intervention approaches.

By the same token, a clear depiction of street gangs facilitates our ability to spot copycat groups and tagger crews as different from street gangs yet as potential precursors of new street gangs. The transition from the one to the other highlights factors likely to increase group cohesiveness, and we can see better how variability in cohesiveness relates to gang duration.

The proliferation issue is perhaps the most salient and disturbing in the current gang era. From a few dozen gang cities in the 1950s, we have progressed at an accelerating rate to more than one thousand (and counting). More and more young lives are getting less and less satisfying in youth groups, and others are more in danger of serious harm and even death. Thousands of gang-related deaths cannot lightly be set aside, especially if we are inadvertently contributing to them by our very attempts to intervene. Furthermore, as by-products of a growing urban underclass and our own propensity to disseminate and commercialize questionable aspects of youth culture, street gangs are telling us that the cores of our towns and cities, and increasing components of our suburbs as well, are deteriorating. Deterioration is not just a matter of abandoned buildings; it also means abandoned lives and abandoned categories of people.

The third theme of gang cohesiveness, beyond its "only academic" concern, is critical to our attempts at direct gang intervention. Gang dynamics have a life of their own; they twist and turn what the intervenors do into sources of greater group bonding. Whether it's the response to the attention of gang workers, the surveillance of police, or the antigang messages of our schools and media, rather uniformly we find gang structures using us to feed themselves. Attacking gangs indiscriminantly, without carefully attending to the cohesiveness-building properties of the attacks, is counterproductive.

The gang and drug issue is perhaps the most temporary of the four, although I've aimed some of my sharpest barbs at those who stress the connection between the two. Lonely in our first attempts to question the gang–crack sales nexus, we now see more and more of the research and police communities becoming aware of the distinction. Maybe federal funding sources and state and federal enforcement agencies can be convinced to take a second look as well. The National Institute of Justice seems to be doing so. The National Academy of Sciences panel on gangs had a chance to do so but dropped the ball. The Administration on Children, Youth, and Families is supporting a wide assortment of gang prevention projects but, by congressional demand, is using drug money to do it. This infects their programs in misleading ways. This, too, I think may soon be changing.

As for the enforcers—the FBI, DEA, ATF, and some of their state-level counterparts—I have yet to see their recognition of the limited relationship between drug distribution and street gangs. They seem so committed to the conspiratorial base of federal intrusion that I doubt their educability. Better that they just drop the street gang angle and continue to do what they do best in the realm of drug distribution. The media, too, have been resistant to modifying their hyperbole about the great gang–drug story. It's truly a topic that exemplifies the way that gang definitions, proliferation, and cohesiveness can combine to confuse and obfuscate where clarity is most needed.

What would I like to suggest as a parting comment in each of the four issues? Try these:

- On gangs and drugs, forget it: enough said.
- On cohesiveness, let's not lose the principles under pressure. In 1973, the then LAPD Assistant Chief Daryl Gates sent an operations management paper to the troops, telling them about the cohesiveness issue and warning them against any actions that would serve to increase gang cohesiveness and "the reinforcement of gang identity."

In the late 1980s, then Chief Daryl Gates declared a war on gangs, approved CRASH and Operation Hammer, and announced that the LAPD would "obliterate" gangs. The earlier principle was the better one, sir.

- On definitions, I think the best, next move would be to undertake a serious program of research designed to document carefully the categories of street gangs and related groups in our communities—street, drug, traditional, skinhead, spontaneous, whatever—for a better understanding of what each brings with it. A decent typology of gangs in the 1990s could help us decide how best to approach them for both prevention and control. Know your gangs.

- On proliferation, my hypothesis about the effect of our exacerbating underclass problems remains salient in my mind. It is even more true now than it was some years ago when one of my street gang informants told me, "This is a white man's world, you know. They got all the good jobs and things, you know. If we went to school it probably wouldn't help no way. I know a lotta colored dudes, they got they high school education, and they can't get a job."

In the long run, it won't be gang intervention programs that reduce gang presence and activity. It will be our success in reclaiming communities with and for their residents. If we're a long way from there now, at least gang proliferation and crime are highlighting the need.

All right, so what can be done? I'm obviously fairly skeptical about the general directions we have taken to gang intervention—poorly conceptualized programs with negligible, sometimes even harmful, effects. General prescriptions are doubly difficult, because short-term solutions are unlikely to be much more than palliative, and longer-term approaches, of the sort needed, may well be beyond our capacities.

Let us look first at the short term. Most forms of detached worker programs have shown the capacity to connect with seriously alienated as well as fringe gang members and, in many instances, have made gains in job finding, tutoring, family intervention, counseling, and the like. But in doing so, they have not demonstrated a consequent crime reduction on an individual level. And worse, they have demonstrated the inadvertent effect of increasing gang cohesiveness and gang crime. If we're going to continue to advocate detached worker programs, let's at least ask them to be area or neighborhood workers, intervening in the lives of any needy youth regardless of gang affiliation and without attempting naively to "use" the gang structure. Kids need help; gang kids may need more, but we must avoid those forms of intervention that will reinforce gang identity.

Suppression programs also are short term. They do little to nothing about the forces that foster gang development. But by focusing on gang-qua-gang problems, they can and—I'm convinced—do serve to reinforce gang identity. Gang members who commit crimes can be arrested. Their gang status doesn't really change that. Individual persons can be handled by the justice system for their individual acts. There's no need to emphasize the gang affiliation. Legislation that increases penalties for gang-related crimes makes prosecutors happier, but it leads cops to play the gang game with even more vigor. It's counterproductive. Perhaps a really well articulated deterrence program could bring some short-term effects. STEP acts come closest to this but, of course, have not been used as tests of their underlying principles.

If we move toward midrange programs, more potential emerges, but so do more problems. Gangs are by-products of communities, and some communities may indeed be able to apply the brakes to gang activity. Community surveillance, parent clubs and parent training, more and more meaningful local job opportunities, more appropriate and sensitive school systems, greater youth–adult contact systems, and the like might provide some local, informal social control with greater power than dependence on patrol cars and park closings. But getting a neighborhood geared up to do this, a neighborhood already shown to be a gang-spawning stream, truly tests our skills in social change. It's worth the effort, and I encourage it, but with limited expectations.

The true solutions, of course, are in the long run. Yet who among us truly believes that our society is willing to make the moves to sterilize the fertile soil from which spring gangs, homelessness, drug abuse, violence, and a host of associated social ills. We'd have to learn to remove institutional racism, to reconstitute a sense of local community, to reduce youthful unemployment and inadequate education and adult unemployment and other sources of marital strain, and otherwise to reform a society that prefers to ignore its darker side. "Power to the people" was the rallying cry of the 1960s generation. In a pluralistic democracy, power must be widely shared to be effective, and so must social responsibility. Failure to share both social power and social responsibility yields a surfeit of social ills. Street gangs are one of these; we don't effectively eliminate gang problems with short-term solutions because we cannot affect their causes in that way. The reader must decide, as each citizen must, what he or she is willing to sacrifice in order to get over gangs and their fellow ills.

As I complete this manuscript, I realize that there are a number of experiences I have not shared with the reader, personal experiences that may have little value to the reader but have shaped my approach to the street gang. Gangs gave me my first whiff of marijuana and my look at youths totally strung out on heroin. They showed me, first hand, that so-called senseless gang violence is not senseless at all; it derives its meaning from hurt, from ego challenges, from exaggerated and manipulated senses of group loyalty. The gang experience taught me to listen to gang members and cops with equal interest, and equal skepticism.

I recall vividly watching "scrunching" at gang dances—scrunching was the physically intimate, rhythmic melding of young bodies that made even chaperoning adults guard their groins—and I remain dismayed at the 16-year-old mothers of the gang world. I watched the male disdain for female integrity but also the female acquiescence to male hegemony. This was no world for the feminist.

The phrase *institutional racism* loses its sterile, disinterested character when one talks to and, more important, listens to young blacks and Hispanics who have already given up at age 16. A better writer than I might have infused these pages with a stronger sense of the futility of the barrio, the set, the 'hood. Street gangs are an amalgam of racism, of urban underclass poverty, of minority and youth cultures, of fatalism in the face of rampant deprivation, of political insensitivity, and the gross ignorance of inner-city (and inner-town) America on the part of most of us who don't have to survive there. I've lived there, but I've never had to survive there. Street gangs are understandable once one has some understanding of the ignored sectors of our society.

None of this excuses street gang crime or violence; instead, it says that gang crime and violence can be understood. Once understood, they might—just might—be alleviated. If so, we all—all of us—be benefited.

Notes

Chapter 1

1. In a late development, that officer's testimony was withdrawn because he was already under "administrative investigation" in his department. The prosecution's intent of seeking sentence enhancements bowed to administrative niceties unrelated to the merits of the case.

2. The Office of Juvenile Justice and Delinquency Prevention (OJJDP) is the principal source of support.

3. The mentality here is reminiscent of Robert Carter's interviews with juvenile officers in Los Angeles. One of them derides academic theories of crime and delinquency, maintaining that only street cops can appreciate what causes crimes: "I've heard all the theories of crime. Let me tell you, crime is caused by assholes. That's the asshole theory." Quoted in Robert M. Carter, "The Police View of the Juvenile Justice System," in *The Juvenile Justice System*, ed. Malcolm W. Klein (Newbury Park, Calif.: Sage, 1976), p. 124.

4. California State Task Force on Gangs and Drugs, *Final Report* (Sacramento: California Council on Criminal Justice, 1989), p. viii.

5. *Delinquency and Society* (Englewood Cliffs, N.J.: Prentice-Hall, 1990), p. xi.

6. An excellent text is that by Herbert C. Covey, Scott Menard, and Robert J. Franzese, *Juvenile Gangs* (Chicago: Thomas, 1992). George Knox produced a massive text for criminal justice programs, *An Introduction to Gangs* (Berrien Springs, Mich.: Van de Vere, 1991); and Arnold Goldstein wrote another aimed at psychology students, *Delinquent Gangs: A Psychological Perspective* (Champaign, Ill.: Research Press, 1991). Equally important is Irving Spergel's forthcoming review of the literature, as well as his chapter, "Youth Gangs: Continuity and Change," in *Crime and Justice: An Annual Review*, vol. 12, ed. Michael Tonry and Norval Morris (Chicago: University of Chicago Press, 1990), pp. 171–267.

7. Throughout the book, I'll use synonyms for gangs, gang members, and assorted others.

8. The Youth Development and Delinquency Prevention Administration.

9. Some years later I learned the proper phrase for this process from Bill

Davidson at Michigan State University: MILTFP41, or Make it like the friggin' picture, for once!

10. Twenty-five years later, that old couch is still in my office. Either they don't make them that way anymore, or my body has gotten insensitive.

11. Malcolm W. Klein, *Street Gangs and Street Workers* (Englewood Cliffs, N.J.: Prentice-Hall, 1971).

12. Our findings have since been replicated and validated on a statewide level by Gary Bailey and Prabha Unnithan of Colorado State University, as reported at the 1993 meeting of the American Society of Criminology in Phoenix.

13. This finding was based on the national study "Gang Prosecution Prosecutor Survey," carried out by the Institute for Law and Justice, Alexandria, Virginia, and reported in June 1992 to the National Institute of Justice.

14. But other departments, including the sheriff's, wouldn't go along with the proposal. Currently, the statistics are being recorded by both definitional systems.

15. Knox, *An Introduction to Gangs*, p. 343.

16. We lucked out here. The Southern California Injury Prevention Research Center at the University of California at Los Angeles was preparing a large proposal for the CDC and invited our participation. Because it was funded, we were funded on its coattails. Thanks, Bruins.

Chapter 2

1. C. Ronald Huff, *Gangs in America* (Newbury Park, Calif.: Sage, 1990).

2. Bill Buford, *Among the Thugs: The Experience, and the Seduction, of Crowd Violence* (New York: Norton, 1991). See also the more careful empirical studies of Belgian soccer hooligans by the research team at the Catholic University of Leuven: K. Van Limbergen, C. Colaers, and L. Walgrave, "The societal and psycho-sociological background of football hooliganism," *Current Psychology Research and Reviews* 8 (Spring 1989): 4–14; and L. Walgrave, "After Heysel," in *Proceedings of the European Conference on Football Violence*, ed. T. O'Brien, Preston Lancashire Polytechnic, 1987. They find these soccer fan groups to approach Yablonsky's concept of "near-groups," that is, socioeconomically and psychologically impaired young men: "There is no specific organization, there are different subgroups that only meet on the terrace. We did not observe accepted leaders, but certain individuals are strongly identified by their heroic courage" (Van Limbergen et al., p. 8).
A somewhat different view, relying heavily on verbatim reports from English soccer hooligans, is provided by Dick Hobbs and David Robins in "The Boy Done Good: Football Violence, Changes and Continuities," *Sociological Review* 39 (1991): 551–79. The article also provides the etymology of the term *hooligan*, which comes from a nineteenth-century British criminal by the name of Patrick Hooligan. The description suggests that Mr. Hooligan would have felt much at home in the soccer riots, but quite out of place in the duller setting of the typical street gang.

3. The captain was hardly alone in his simplistic "thugs and hoodlums" definition. David Dawley, a white activist working with Chicago's black gang,

the Vice Lords, quotes a local official there as follows: The state attorney for Cook County, Edward Hanrehan: "All I see gangs doing is marauding, ravaging, shooting and killing."

One wonders how comfortable these enforcement officials would be with some of their definitional company. The Study Committee of the National People's Congress of mainland China offers the most severe penalties for "hoodlum cliques," or gangs. They are described as "the dregs of society" who "hurt and kill people ruthlessly and undermine social rules." This was reported in the *China Legal Daily*, June 1984, by Cui Nanshan.

4. In an infinite regression of ambiguity, Walter Miller decries even this approach. In a consultant statement offered to Spergel's gang suppression project, Miller doubted our ability to define even *street* successfully.

5. The latest work on skinheads, and a very useful one, is Mark S. Hamm's *American Skinheads: The Criminology and Control of Hate Crime* (Westport, Conn.: Praeger, 1993).

6. Howard Pinderhughes, "The Anatomy of Racially Motivated Violence in New York City: A Case Study of Youth in Southern Brooklyn," *Social Problems* 40 (November 1993):478–92.

7. An amusing depiction of the bikers' lifestyle is provided by Rene Becker's article, "Gangbusters," in the *New Republic*, October 27, 1979. The amusement is at the expense of a congressional committee looking at the bizarre behaviors of bikers, under the guise of welfare reform. The most recent treatise on the bikers as a very different group is David R. Wolf's *The Rebels* (Toronto: University of Toronto Press, 1991).

8. Sergeant Bob Jackson (Los Angeles Police Department) and Sergeant Wes McBride (Los Angeles Sheriff's Department), two longtime police gang experts, wrote a useful practitioner's text that includes material on stoner groups. They are more ambivalent than I am about not classing stoners as street gangs. There's also a nice *New Yorker* article, "Reporter at Large: The Crazy Life," by Bill Barich in the November 3, 1986, issue.

9. Jackson and McBride provide some useful materials on this issue as well as on the stoners.

10. Almost twenty-five years ago, Bernard Cohen showed that gangs were significantly more homogeneous on a number of factors than were other delinquent groups. Despite some methodological problems, Cohen's work still deserves more attention than it has received. See his two chapters in Thorstein Sellin and Marvin Wolfgang, eds., *Delinquency: Selected Studies* (New York: Wiley, 1969). Spergel also argues persuasively for distinguishing between gangs and other youth groups that occasionally become involved in delinquent episodes; see his *Youth Gangs: Problems and Response* (New York: Oxford University Press, forthcoming). As of this writing, Spergel's *Youth Gangs* has not been released by Oxford University Press, but the author has generously given me permission to use materials from it in this book.

Miller found street gangs to be a rather small proportion of all "law-violating" groups around the country. He is quoted sardonically by Alfredo Gonzales in his dissertation (University of California at Berkeley) in the case of the Philadelphia police who claimed to find only 88 gangs in their city, along with many other groups that didn't meet their gang criteria. Yet when the department needed additional funds from the city fathers, they suddenly

found an estimated 237 gangs! The problem, of course, is that in the absence of a clear agreement on the meaning of the term *gangs*, and therefore the phenomenon, it is subject to all kinds of social construction to meet the practical needs of politics, media sales, and funding (or just the opposite, as in the gang member's denial, "We a defensive club, man."). See Miller's 1979 report to the Office of Juvenile Justice and Delinquency Prevention, *Crime by Youth Groups and Gangs in American Cities*.
Recent descriptions of "punks" in Canada were given to me by Stephen Baron at the University of Victoria. See his "Resistance and Its Consequences: The Street Culture of Punks," in *Youth and Society* 21 (1989): 207–37, and "The Canadian West Coast Punk Subculture: A Field Study," *Canadian Journal of Sociology* 14 (1989): 289–316.

11. George W. Knox, *An Introduction to Gangs* (Berrien Springs, Mich.: Van de Vere, 1991), p. 6.

12. Delbert S. Elliott and Scott Menard, "Delinquent Friends and Delinquent Behavior: Temporal and Developmental Patterns" (Boulder: University of Colorado, Institute of Behavioral Science, undated).

13. See Finn-Aage Esbensen and David Huizinga, "Gangs, Drugs, and Delinquency in a Survey of Urban Youth," *Criminology* 31 (November 1993): 565–87.

14. Terrence P. Thornberry et al., "The Role of Juvenile Gangs in Facilitating Delinquent Behavior," *Journal of Research in Crime and Delinquency* 1 (1983): 55–87.

15. Merry Morash, "Gangs, Groups, and Delinquency," *British Journal of Criminology* 4 (1983): 316, by permission of the Oxford University Press. I guess now she'll know who that anonymous critical reviewer was.

16. Robert J. Bursik and Harold G. Grasmick, *Neighborhoods and Crime: The Dimensions of Effective Community Control* (Lexington, Mass.: Lexington Books, 1993), p. 123.

17. Walter B. Miller, "American Youth Gangs: Past and Present," in *Current Perspectives on Criminal Behavior*, ed. A. S. Blumberg, 2nd ed. (New York: Knopf, 1981), p. 292.

18. Dr. Lawrence Breen and Sergeant Martin Allen, "Gang Behavior: Psychological and Law Enforcement Implications," p. 20, in the February 1983 issue.

19. I was shown by another writer a research proposal submitted to a private foundation that sought support for a study of incarcerated members of "violent gangs." The foundation supports research on violence, so perhaps the proposal writer might be excused for using the term *violent gang*, but not for the persistence of its theme throughout the proposal. The impression is left that youth gangs—and their members—are characterized by violence, but supportive data were not provided and contrary data were not discussed.

20. As fine a description and analysis of contrasting youth groups as can be found is given by Mercer L. Sullivan in *Getting Paid: Youth, Crime, and Work in the Inner City* (Ithaca, N.Y.: Cornell University Press, 1989).

21. There's an interesting side issue here. In several studies, including the very carefully crafted longitudinal studies in Denver and Rochester, a self-nominated criterion of gang membership was used, along with some purifying by delinquency involvement. If a youngster answered yes to the question of whether he was a gang member, this criterion, applied along with one of

some criminal involvement, could be used as the operational definition of street gang membership. Admittedly, such a definition, dependent solely on the youth's self-report, seems questionable. Yet when the self-nominated members and nonmembers were compared on a number of items, they turned out to be distinctively different. This makes the researcher's job much easier, at the very least. The same approach employed with New Mexico gangs was contrasted with a membership criterion based on the youth's behavioral indicators—initiations, colors, tattoos, handsigns, use of graffiti, and so on. These researchers found that the difference in several variables by the self-nomination was considerably greater than that in the use of the behavioral indicators of gang membership. Thus the evidence mounts that street gang members do indeed come to view themselves as such and as distinct from other groups. See L. Thomas Winfree Jr., Kathy Fuller, Teresa Vigil, and G. Larry Mays, "The Definition and Measurement of 'Gang Status': Policy Implications for Juvenile Justice," *Juvenile and Family Court Journal* 43 (1992): 29–37.

22. Scott Decker, "Gangs and Violence: The Expressive Character of Collective Involvement" (unpublished paper, University of Missouri at St. Louis, undated). Cited by permission of the author.

23. Joan Moore, *Going Down to the Barrio: Homeboys and Homegirls in Change* (Philadelphia: Temple University Press, 1992), p. 132.

24. The Maxson–Klein national gang migration survey, to be discussed later, revealed that the most common reason for what gang migration does exist is family and residential moves, not the much ballyhooed trend toward drug sales franchising. That the migration of gang members, though not unusual, is not to blame for the emergence of gangs in new cities was nicely documented for Milwaukee by John Hagedorn, *People and Folks: Gangs, Crime, and the Underclass in a Rustbelt City* (Chicago: Lakeview Press, 1988); and for Kenosha, Wisconsin, by Richard Gerity and Susan Tanaka, "Metropolitan Gang Influence and the Emergence of Group Delinquency in a Regional Community," *Journal of Criminal Justice* 20 (1992): 93–106. Both cities are often thought to be infiltrated by Chicago gangs, but their problems are homegrown.

25. In our ongoing survey of 1,100 cities suspected of having gangs or experiencing gang migration, 792 reported one or more street gangs. The 1,100 was a nominations base, not a known gang city base. It suggests that the estimate of 1,000 gang cities is in no way out of line.

26. This quotation is not verbatim, but as close as I can come from my own notes taken after the conversation.

27. From John M. Hagedorn, "How Do Gangs Get Organized?" in *Juvenile Delinquency: A Justice Perspective*, ed. R. A. Weisheit and R. G. Culbertsen, 2nd ed. (Prospect Heights, Ill.: Waveland Press, 1990) p. 150.

28. Scott Decker and Barrik Van Winkle, "Slinging Dope: The Role of Gangs and Gang Members in Drug Sales" (draft paper, University of St. Louis at Missouri, September 1993).

29. For readers interested in descriptions of typical gang worker programs, I recommend the New York City Youth Board's *Reaching the Fighting Gang*; Frank Carney, Hans W. Mattick, and John D. Callaway, *Action on the Streets* (New York: Association Press, 1969); and Irving I. Spergel's *Street Gang Work: Theory and Practice* (Reading, Mass.: Addison-Wesley, 1966). Data on worker

activities are laid out in my *Street Gangs and Street Workers* (Englewood Cliffs, N.J.: Prentice-Hall, 1971).

30. These data and the others summarized here are fully reported in my *Street Gangs and Street Workers*.

31. Examples of low-companion offenses in our data were disturbing the peace, most status offenses, and malicious mischief. Examples of high-companion offenses were assault, auto theft, and curfew violation.

32. One cautionary note is required. Leon Jansyn studied one small gang in Chicago. The decline and increase in cohesiveness and delinquency could have been a matter of several variables in this one-year analysis, but he found that when the gang's cohesion dropped to quite a low point, the criminal activity increased. His reasoning was that this increase was a deliberate undertaking to reassert the gang's strength. It is a perfectly reasonable hypothesis, complementing but not necessarily opposing mine. However, it needs to be validated and has not been, to my knowledge. See Leon Jansyn, "Solidarity and Delinquency in a Street Corner Group," *American Sociological Review* 31 (1966): 600–14.

Chapter 3

1. "You have often heard of me. Now take a good look. I am Imaj no Shiro Kamehira.... As I am a valiant warrior among the men of Lord Yoshinaka, your master, Yoritomo at Kamakura must know my name well. Take my head and show it to him." Kamehira then felled the next eight enemy warriors with his next eight arrows (p. 22).
Other examples can be found in the translation by Hiroshi Kitagawa and Bruce T. Tsuchida, published by the University of Tokyo Press, 1975.

2. The reader less bothered by definitional issues will find a comprehensive review of gangs throughout history in Chapter 5 of Herbert C. Covey, Scott Menard, and Robert J. Franzese's *Juvenile Gangs* (Chicago: Thomas, 1992).

3. Gilbert Geis, "Juvenile Gangs" (paper prepared for the President's Committee on Juvenile Delinquency and Youth Crime, June 1965), p. 1.

4. The reader is referred to the texts listed in the Notes.

5. Albert K. Cohen, *Delinquent Boys: The Culture of the Gang* (New York: Free Press, 1955); and Richard A. Cloward and Lloyd E. Ohlin, *Delinquency and Opportunity: A Theory of Delinquent Gangs* (New York: Free Press, 1960).

6. James F. Short and Fred L. Strodtbeck, *Group Process and Gang Delinquency* (Chicago: University of Chicago Press, 1965).

7. But for a prime example to the contrary, see (if you must) Charles Patrick Ewing's *When Children Kill: The Dynamics of Juvenile Homicide* (Lexington, Mass.: Lexington Books, 1990), chap. 6, on gang killings.

8. See J. Michael Olivero, *Honor, Violence, and Upward Mobility: A Case Study of Chicago Gangs During the 1970s and 1980s* (Edinburg, Tex.: Pan American Press, 1991), as well as the description provided by George Knox, in *An Introduction to Gangs* (Berrien Springs, Mich.: Van de Vere, 1991).

9. Some refer to the Los Angeles Crips and Bloods as similar organizations, but this is not true. The internecine fighting among Crip gangs and among Blood gangs is symptomatic of the irrelevance of supergang thinking in Los Angeles.

10. "Chicano Gangs: A History of Violence," *Los Angeles Times*, December 11, 1988, p. 1.

11. Again, see the structural depiction in my *Street Gangs and Street Workers* (Englewood Cliffs,N.J.: Prentice-Hall, 1971). An independent confirmation, using block-modeling techniques, was provided by Phipps Arabie of the Bell Laboratories and the University of Illinois. The clique structures emerged in his analysis of my data, prepared under an LEAA (Law Enforcement Assistance Administration) grant in 1981.

12. Walter B. Miller, "American Youth Gangs: Past and Present," in *Current Perspectives on Criminal Behavior*, cd. A. S. Blumberg, 2nd ed. (New York: Knopf, 1981) pp. 297–98.

13. See Gay Luce, "Delinquent Girl Gangs," in *Mental Health of the Child*, ed. Julius Segal (Rockville, Md: National Institute of Mental Health, 1971); Lee H. Bowker, Helen Shimota Gross, and Malcolm W. Klein, "Female Participation in Delinquent Gang Activities," *Adolescence* 15 (1980): 509–19.

14. Anne Campbell, *The Girls in the Gang*, 2nd ed. (Cambridge, Mass: Basil Blackwell, 1991), p. 277.

15. Joan Moore documents a few exceptions, as reported to her, from the 1930s to the 1950s. See her *Going Down to the Barrio* (Philadelphia: Temple University Press, 1991).

16. Michael McAleenan, "Sex-Role Behavior in an Integrated Male/Female Gang" (draft paper, Department of Sociology, Occidental College, 1974).

17. See Moore, *Going Down to the Barrio*.

18. The point is nicely made by William Smart in his "Female Gang Delinquency: A Search for `Acceptably Deviant Behavior'," *Mid-American Review of Sociology* 15 (1991): 43–52.

19. For instance, Miller showed a range, across fourteen gangs, of court-recorded violent crimes, from 1.2 per member to 8.3 per member. See Walter Miller, "Violent Crimes in City Gangs," *Annals of the American Academy of Political and Social Science* 364 (1966): 96–112; Moore, Vigil, and Garcia report cliques in the 1950s and 1960s, ranging from 11 to 118 members. See Joan Moore, Diego Vigil, and Robert Garcia, "Residence and Territoriality in Chicano Gangs," *Social Problems* 31 (1983): 182–94; Hubbard and Cooper, in the Chicago YMCA gang project, recorded cliques whose youngest members ranged from age 12 to age 18 and cliques whose oldest members were as young as 15 and as old as 32. See Fred D. Hubbard and Charles N. Cooper, "Program for Detached Workers: Progress Report" (mimeo), Chicago YMCA, March 1963.

20. Miller, "Violent Crimes in City Gangs," p. 110.

21. Quoted in James E. McCarthy and Joseph S. Barbaro, "Redirecting Teen-age Gangs," in *Reaching the Unreached*, ed. S. Furman (New York: New York City Youth Board, 1952), p. 113.

22. Short and Strodtbeck, *Group Process and Gang Delinquency*.

23. Miller, "American Youth Gangs," p. 293.

24. Walter B. Miller, "White Gangs," *Trans-action*, September 1969, p. 26.

25. The data come from the commission's final report, *Gang Violence in Philadelphia*.

26. Bernard Cohen, "The Delinquency of Gangs and Spontaneous Groups," in *Delinquency: Selected Studies*, ed. Thorsten Sellin and Marvin E. Wolfgang (New York: Wiley, 1969).

27. Taken from Table 5 in Walter B. Miller's "Violent Crimes in City Gangs."

28. Irving Spergel, personal communication, January 1984.

29. It's hard to be a gang researcher these days without also being strongly in favor of gun control. I mean serious gun control, not just delaying tactics. As one young man put it, "They put a hold on a gun for fifteen days in the gun shop, and the shootin' is on day 16. They ain't no rusty guns out there." One gang member, known to have shot, although not killed, his stepfather at an early age, bragged about having his gun on him on a number of occasions. He was easy to get to know, as I did over some months, so on one of these gun-bearing occasions I had a long conversation with him, offering to take and keep the weapon or destroy it, using my best rationalizations about danger to self and others. I made absolutely no headway, but the following week he sauntered over to me saying, "Hey Doc, you remember that gun you wanted me to get rid of?" I said I did, and he said, "Yeah, well I did it." In the midst of my heartfelt commendation and verbal rewards (and self-satisfaction, I must admit), he interposed, "Yeah, see, I traded it in for a Luger." To this day, I don't know if he was putting me on or in fact made a trade. But there was no mistaking his pride of ownership.

30. Diego Vigil, "The Established Gangs," in *Gangs: The Origins and Impact of Contemporary Youth Gangs in the United States*, ed. Scott Cummings and David J. Monti (Albany: State University of New York, 1993), pp. 99–100.

31. Vincent Riccio, as told to Bill Slocum, "My Life with Juvenile Gangs," *Saturday Evening Post*, September 15, 1962.

32. Interestingly, Miller saw the aggression as due to the absence of fathers, producing a female-dominated household that required aggression to reassert one's manhood, whereas Moore laid more stress on the inadequate and "grouchy" father (see Joan Moore, *Going Down to the Barrio*).

33. Carl C. Werthman, "Delinquency and Authority" (master's thesis, University of California at Berkeley, 1964), p. 25.

34. C. J. Friedman, F. Mann, and A. S. Friedman, "A Profile of Juvenile Street Gang Members," *Adolescence* 10 (1975): 563–607.

35. From an interview with Janice Castro in *Time*, March 16, 1992.

36. Excerpted with Dr. Short's permission from his June 11, 1980, public lecture, "Youth Gangs: Images and Realities," part of the Boys' Town Center Lecture Series on Children and Families.

37. "Delinquency Theory and Recent Research," *Journal of Research in Crime and Delinquency* 4 (1967): 35. This was a special issue devoted to gang research, edited by James F. Short Jr.

38. "The Organization and Behavior of Violent Gangs" (paper delivered at the American Academy for the Advancement of Science meetings in New York, December 27, 1960).

39. The tentative truce between some Crip and Blood sets in postriot Los Angeles is holding weakly as I write this. Not originally adult sponsored, this was an indigenous movement among the gang members, which may explain its initial success. Stay tuned.

40. An example of such programs' disastrous consequences is provided by Richard W. Poston in *The Gang and the Establishment* (New York: Harper, 1991). This concern resurfaced recently and in particular in the writing of Martin Sanchez-Jankowski, *Islands in the Street: Gangs and American Urban Society* (Berkeley and Los Angeles: University of California Press, 1993).

41. Cited in Lewis Yablonsky, "The Organization and Behavior of Violent Gangs," paper presented at the American Academy for the Advancement of Science, December 1960, p. 6.

42. Controversy remains over the facts and myths of supergang involvement in these and other enterprises. Irving Spergel, in a private communication, suggested that the involvement was often quite indirect.

43. Very recently, in the aftermath of the postverdict Los Angeles riots, a few older gang members formed a group called Hands Across Watts. Concerned with establishing peace among gangs, the leadership was labeled both courageous for their attempt and dangerous in view of their past gang crime records. In a public presentation to high school youths considering undertaking community action projects for gang control, I offered some background on gang matters and the elements of community control. I was followed by the two founders of Hands Across Watts, who declared that all I had to say was "bullshit," the ravings of a white university professor and therefore worthless. The maps of gang proliferation (see Chapter 4) were called *garbage* by one founder and *less than garbage* by the second.
The point is not the validity of what I might have offered but the continuing narrow perspective of committed gang members. Peace or no peace, they were still thoroughly encapsulated in the gang culture. A few months later, on January 13, 1992, while participating in an evening of gambling, one of the two died in a hail of bullets, despite his habit of wearing a bullet-proof vest. Committed gangbangers, no matter how much they profess being reformed, tend to maintain lifestyles that raise dramatically their chances for criminal involvement, be it as perpetrator or as victim. In the case of Tyrone "Tony Bogard" Thomas of Hands Across Watts, it was both.

44. See Irving Spergel, "Youth Gangs and Urban Riots," in *Riots and Rebellion: Civil Violence in the Urban Community*, ed. L. H. Masotti and D. R. Bower (Beverly Hills, Calif.: Sage, 1968), p. 143.

45. Short, "Delinquency Theory and Recent Research," p. 3.

46. See Martin Sanchez-Jankowski's previously cited *Islands in the Streets*.

Chapter 4

1. Consider the following:
- Of twenty-nine gangs named in these inner-city interviews, only two were named four times, and only three were named twice. Thus twenty-four were named only once. This lack of consensus suggests little acknowledged gang presence.
- Of initiation procedures, fighting received three mentions. A fourth was truly whimsical; "Get five different girls' phone numbers and prove them within one hour."
- Of weapons, guns were mentioned only three times, the rest being knives, bottles, bricks, bats, sticks, and stones.
- A total of nineteen gang neighborhoods were listed, but none more than three times and most only once or twice.
- Finally, gang activities elicited a much more appropriate, cafeteria list of offense categories but also some very unganglike responses: Build a shack in the woods, go to movies and go out with girls, get money without paying taxes, scare people for fun, misuse girls, and play hoops.

Clearly, we're some distance from committed street gang membership here (as opposed to Pittsburgh's other claim to gang fame, its drug gangs). Needless to say, I'm indebted to Rolf and Magda Loeber for sharing these data with me. I received an invitation, quite independent of all this, to travel to Pittsburgh to give several lectures on gangs and to consult with social agencies there because (1) either the city had a gang problem or (2) it expected to have one. It was, under the circumstances, an irresistible invitation. I used my two days in the city as much as an investigator as a consultant.

2. Evidence was offered by John W. C. Johnstone in 1981, in suburban Chicago areas; by Stapleton and Needle in 1982, who found twenty-seven gang cities in a sample of sixty; by Rosenbaum and Grant in 1983, on the evolution of gangs in Evanston, Illinois; by Walter Miller's mid-1970s national survey materials which, though unpublished, were not unknown to gang scholars; and even by Cartwright and Howard as early as 1966 and Hardman in 1969. Johnstone pointed out, "Suburban gangs are less visible than their inner-city counterparts, in part, perhaps, because the metropolitan media pay less attention to them" (p. 372). See John W. C. Johnstone, "Youth Gangs and Black Suburbs," *Pacific Sociological Review* 24 (1981): 355–575; William Vaughan Stapleton and Jerome A. Needle, *Response Strategies to Youth Gang Activity* (Sacramento, Calif.: American Justice Institute, 1982); Dennis P. Rosenbaum and June A. Grant, *Gangs and Youth Problems in Evanston: Research Findings and Policy Options* (Evanston, Ill.: Northwestern University Center for Urban Affairs and Policy Research, 1983); Walter B. Miller, *Violence by Youth Gangs and Youth Groups as a Crime Problem in Major American Cities* (Washington D.C.: U.S. Department of Justice, 1975); Desmond S. Cartwright and Kenneth J. Howard, "Multivariate Analysis of Gang Delinquency: I. Ecologic Influences," *Multivariate Behavioral Research I*, July 1966, pp. 321–71; and Dale G. Hardman, "Small Town Gangs," *Journal of Criminal Law, Criminology and Political Science* 2 (1969): 173–81.

3. For example, Houston, despite its size, remained free of gangs. Then one day Jim Short called (he was on leave there as a visiting professor). "It's OK, Mac, Houston finally has street gangs." Another city had fallen. The Houston police department reported 17 gangs in 1988, 56 in 1989, 83 in 1990, and more than 100 in 1991.

4. See his "The Rumble This Time," *Psychology Today*, May 1977, p. 52.

5. I am indebted to my geographer colleague Curt Roseman and to Cheryl Maxson for their work on this material. We hope to continue in the future to make to more sophisticated analyses with these combined files. Some cities are missing because of the respondents' inability to specify when gangs first appeared in their jurisdictions.

6. Daniel J. Monti, "Gangs in More- and Less-Settled Communities," in *Gangs: The Origins and Impact of Contemporary Youth Gangs in the United States*, ed. Scott Cummings and David J. Monti (Albany: State University of New York, 1993), pp. 219–53. The variations described by Monti, those clearly discernible to residents of the Los Angeles area or often described in Chicago, New York, San Francisco, and other metropolitan areas, are mirrors of the national variations described in 1971 in my *Street Gangs and Street Workers* (Englewood Cliffs, N.J.: Prentice-Hall) and more recently by Fagan and by Sanchez-Jankowski in their multicity studies. See Jeffrey Fagan, "The Social Organization of Drug Use and Dealing Among Urban Gangs," *Criminology* 27

(1989): 633–69; and Martin Sanchez-Jamkowski, *Islands in the Street: Gangs and American Urban Society* (Berkeley and Los Angeles: University of California Press, 1993).

7. With his permission, I cite here the 1960s-like description of the 1990s street gangs in St. Louis:

Major Findings

Membership Issues: Gang membership primarily includes young males, though female gangs are growing; gangs are very turf oriented, and splinter gangs often form from specific neighborhoods; many gangs emerge from neighborhood friendship groups that begin by pursuing legitimate activities; recruitment is seldom coerced; initiation rituals vary across gangs and often by gang member; most gangs have little formal structure, few key leaders and few internal rules—in short, they are not well-organized; many decisions to join the gang are made in response to violence or the threat of violence from gangs in other neighborhoods; many members leave the gang at will.

Gang Activities: Non-criminal activities outnumber criminal activities; criminal activities (especially non-violent ones) take place in small cliques or subgroups; drug sales are not required of every member and often are not well organized (younger members seldom sell directly); violence is a key feature of gang activity and is expected of all members, occurring both within and between gangs; most gang members report that school is not the primary site of their gang activities.

8. From "Gangs, Drugs, and Violence," in *Drugs and Violence: Causes, Correlates, and Consequences*, ed. M. de la Rosa, E. Y. Lambert, and B. Gropper (Washington, D.C.: National Institute on Drug Abuse, 1990), p. 165.

9. See Ray Hutchinson and Ray Kyle, "Hispanic Street Gangs in the Chicago Public Schools," in *Gangs: The Origins and Impact of Contemporary Youth Gangs in the United States*, ed. S. Cummings and D. Monti (Albany: State University of New York Press, forthcoming); Joan Moore, *Going Down to the Barrio* (Philadelphia: Temple University Press, 1991); Joan W. Moore, "Changing Chicano Gangs," in *ISSR Working Papers on the Social Sciences*, University of California at Los Angeles, 1989–90; John Hagedorn, *People and Folks: Gangs, Crime, and the Underclass in a Rustbelt City* (Chicago: Lakeview Press, 1988); and C. Ronald Huff, "Youth Gangs and Public Policy," *Crime and Delinquency* 35 (1989): 524–37.

10. Commander Robert Dart, quoted in *Midwestern Gang Investigators' Association Report*, November 1992. The Illinois State Police provides a breakdown of the major gangs in the two major "nations" as follows: The Folk includes Black Gangster Disciples, Gangster Disciples, Maniac Latin Disciples, Simon City Royals, and Two Sixers; and The People include Black P. Stone, Cobra Stones, El Rukns, Insane Deuces, Insane Unknowns, Latin Kings, and Vice Lords. See *I.S.P. Criminal Intelligence Bulletin* 49 (April 1992).

11. An explicit exception is provided in James R. Lasley's "Age, Social Context, and Street Gang Membership: Are `Youth' Gangs Becoming `Adult' Gangs?" *Youth and Society* 23 (June 1992): 434–51. However, Lasley's methods do not seem all that appropriate to his test: Even he finds 35 percent of gang membership to be adult, and his upper range is truncated, with no mention of the oldest age.

12. For San Francisco, see Dan Waldorf, "Don't Be Your Own Best Customer—Drug Use of San Francisco Ethnic Gang Drug Sellers," *Crime, Law,*

and Social Change, 1993. I am indebted for the recent Hawaiian data to Meda Chesney-Lind, who shared with me her draft paper "Gangs and Delinquency in Hawaii," prepared for the Hawaii state legislature, November 1992. The Chicago data came from Dart, *Midwesterm Gang Investigators' Report.*

13. Moore reports this pattern in her East Los Angeles gangs in her *Going Down to the Barrio*, p. 47.

14. A useful description of differences in the backgrounds of various Asian gangs and between relevant Asian and Western values can be obtained from the House of Hope in Fresno, California.

15. Carolyn Rebecca Block, "Lethal Violence in the Chicago Latino Community," draft prepared as Chapter 7 in *The Dynamics of the Victim–Offender Interaction*, ed. A. V. Wilson (in preparation).

16. G. David Curry and Irving A. Spergel, "Gang Involvement and Delinquency Among Hispanic and African-American Adolescent Males," *Journal of Research in Crime and Delinquency* 29 (1992): 273–91.

17. These data come from the Maxson–Klein gang migration study and include only those cities that could report on this issue. About two hundred cities could not, certainly not a sign of a strong connection, in any case.

18. Both cities are in Orange County, close to the early Vietnamese resettlement camps. Estimates are that as many as 15 percent of the county's gang members are Vietnamese.

19. The Honolulu situation, as might be expected, is rather unusual. Chesney-Lind's report, "Gangs and Delinquency in Hawaii," cited earlier, shows 47 percent Filipino, 18 percent Samoan, 14 percent Hawaiian or part Hawaiian, and the remaining 21 percent white or other.

20. The U.S. Senate Government Affairs Subcommittee has just released a report of its fifteen-month investigation of Asian gangs. Two comments about the report strike a familiar chord. As noted in the *Los Angeles Times* (December 29, 1992), "The report . . . contained no statistics to support its conclusions but was based on anecdotal evidence supplied by local police and the FBI." And as a correlate to this revealing comment, the *Times* notes that "Federal law enforcement agencies are hampered in their fight against the escalating threat by a lack of foreign language skills, an inadequate knowledge of Asian culture, and a failure to gather and share criminal intelligence."

21. See Ko-lin Chin, *Chinese Subculture and Criminality* (Westport, Conn: Greenwood Press, 1990); also Jeffrey Fagan, Ko-lin Chin, and Robert Kelly, "Lucky Money for Little Brother: The Prevalence and Seriousness of Chinese Gang Extortions" (paper presented at the Multinational Conference on Asian Organized Crime, San Francisco, September 1991). For a contrast, Calvin Toy's description of Chinese gangs in San Francisco sounds very Hispanic/black in style, with their rivalries, need for protection, names, recruitment approaches, need for belonging, status, and identity, and so on. What is still missing is any mention of territoriality. See Calvin Toy, "A Short History of Asian Gangs in San Francisco," *Justice Quarterly* 9 (December 1992): 647–65.

22. Sonni Efron, "Violent, Defiant Vietnamese Form Girl Gangs," *Los Angeles Times*, December 12, 1989, pp. A3, A26.

23. An exception seems to exist in the data from the Rochester Youth Study, in which male–female crime differences are almost nonexistent. However, the members are only in the 13 to 15 age range, far closer to peak

activity for females than for males. The comparison is truncated by the age selection and does not reflect the more active criminal ages. See Beth Bjerregaard and Carolyn Smith, "Gender Differences in Gang Participation and Delinquency," cited with the authors' permission.

24. See Irving Spergel, "Youth Gangs: An Essay Review," *Social Service Review*, March 1992 (draft supplied by author); Joan Moore's *Going Down to the Barrio*; and Anne Campbell's *The Girls in the Gang*, 2nd ed. (Cambridge, Mass.: Basil Blackwell, 1991).

25. See Anne Campbell, "Female Gang Members' Social Representations of Aggression" (1992), an American Society of Criminology paper available from the University of Teeside in England.

26. See David Lauderback, Joy Hansen, and Dan Waldorf, "Sisters Are Doin' It for Themselves: A Black Female Gang in San Francisco," *The Gang Journal* 1 (1992): 57–70. All the other six groups under study were of the auxiliary rather than autonomous type, as is typical.

27. Paul Tracy, *Subcultural Delinquency: A Comparison of the Incidence and Seriousness of Gang and Nongang Member Offensivity* (Philadelphia: University of Pennsylvania, Center for Studies in Criminology and Criminal Law, 1979).

28. These data are taken from and fully covered in Cheryl Maxson, Margaret A. Gordon, and Malcolm W. Klein, "Differences Between Gang and Nongang Homicides," *Criminology* 23 (1985): 209–22.

29. Lawrence J. Bobrowski, *Collecting, Organizing, and Reporting Street Gang Crime* (Chicago: Chicago Police Department, Special Function Group, n.d.).

30. Richard Block and Carolyn Block, *Lethal and Nonlethal Street Gang Crime in Chicago*, prepared for the National Institute of Justice, 1992, and cited with the authors' permission. The four gangs are the Black Gangster Disciples Nation, Latin Disciples, Latin Kings, and Vice Lords.

31. It has become common to blame the increase in gang-related deaths on the use of automatic weapons, assault weapons, and the like. The data available from Los Angeles provide little support. The Sheriff's Department reported a total of 16,155 weapons involved in gang cases during 1992. Of these, 75 percent were firearms of one kind or another. Of these in turn, 75 percent were handguns, 6 percent were shotguns, 3 percent were rifles, and only eight-tenths of 1 percent were assault weapons. Of the firearms actually recovered by the police, 3.5 percent were assault weapons. Although Uzis and AK-47s are highly destructive, they have little to do with gang violence; handguns of various kinds are still the weapons of choice in the hands of gang members.

32. Bureau of Criminal Statistics, *Homicide in California, 1989* (Sacramento: Office of the Attorney General, 1990).

33. See Joan Moore's *Going Down to the Barrio: Homeboys and Homegirls in Change* (Philadelphia: Temple University Press, 1992), and her *Going Down to the Barrio*, cited earlier.

34. H. Range Hutson et al., "Gang Violence: The New Epidemic?" Cited with the authors' permission.

35. I am not pleased to report that in its January 1992 *Street Gang Manual*, the LASD updated the 1991 figure to 795.

36. My wife's comment on hearing this statement: "That's the chief of police? Makes you feel good, doesn't it?"

37. One way to handle the hoodlums, he suggests, is by boot camps, but

"the ACLU would challenge the act of breathing."

38. The statement is more understandable when one looks at the sources of information acknowledged in the commission's report: Forty-two from law enforcement, two from a local university, and one with special thanks to Kelly Services!

39. Cited in James J. Collins, "Summary Thoughts About Drugs and Violence," in *Drugs and Violence: Causes, Correlates, and Consequences*, ed. M. De La Rosa, E. Y. Lambert, and B. Gropper NIDA Research Monograph no. 103, 1990, p. 268. Collins adds, "There is virtually no systematic evidence to support this characterization."

40. Cited in Robert Bursik and Harold Grasmick, *Neighborhoods and Crime: The Dimensions of Effective Community Control* (Lexington, Mass.: Lexington Books, 1993), p. 133.

41. Tyrone F. Price, "Ethnic Succession (Black Gangs)," *Journal of Research on Minority Affairs* 2 (1991): 32.

42. Bruce D. Johnson, Terry Williams, Kojo A. Dei, and Harry Sanabria, "Drug Abuse in the Inner City: Impact on Hard-Drug Users and the Community," in *Drugs and Crime*, ed. M. Tonry and J. Q. Wilson (Chicago: University of Chicago Press, 1990), pp. 21–22.

43. These comments were first developed as part of a proposal to the National Institute of Justice to study gang migration. The proposal, written with Cheryl Maxson, was approved and is now well under way. These comments nonetheless were mine, and Cheryl is absolved of any complicity. The reader wishing to follow Skolnick's argument is referred to three sources. The first two have been published together under single authorship, although my comments and citations apply to the original papers including coauthor attribution: (1) "Gang Organization and Migration" and (2) "Drugs, Gangs, and Law Enforcement" (Sacramento: State of California Department of Justice, 1990); and (3) "Gangs and Crime Old as Time: But Drugs Change Gang Culture," published as a commentary in *Crime and Delinquency in California, 1980–1989* (Sacramento: Department of Justice, 1991). This article contains Skolnick's clearest statement about cultural versus instrumental gangs.

44. Jerome Skolnick, "The Social Structure of Street Drug Dealing," *American Journal of Police* 1 (1990): 5.

45. Quotation is from John C. Quicker, Yvonne Nunley Galeai, and Akil Batani-Khalfani, "Bootstraps or Noose: Drugs in South Central Los Angeles," draft report to the Social Science Research Council, March 1993, p. 36. See also Price, "Ethnic Succession"; and Tracy, *Subcultural Delinquency*.

46. Skolnick, "The Social Structure of Street Drug Dealing," p. 11.

47. Lauderback, Hansen, and Waldorf, "Sisters Are Doin' It for Themselves," p. 148.

48. See James F. Short and Fred L. Strodtbeck, *Group Process and Gang Delinquency* (Chicago: University of Chicago Press, 1965); Irving Spergel, *Street Gang Work: Theory and Practice* (Reading, Mass.: Addison-Wesley, 1966); Frank Carney, Hans W. Mattick, and John D. Callaway, *Action on the Streets* (New York: Association Press, 1969); Malcolm Klein, *Street Gangs and Street Workers* (Englewood Cliffs, N.J.: Prentice-Hall, 1971); John Hagedorn, *People and Folks: Gangs, Crime, and the Underclass in a Rustbelt City* (Chicago: Lakeview Press, 1988); and C. Ronald Huff, ed., *Gangs in America* (Newbury Park, Calif.: Sage, 1989).

49. Skolnick, "The Social Structure of Street Drug Dealing."

50. See Block and Block, *Lethal and Nonlethal Street Gang Crime*, p. 15.

51. Collins, "Summary Thoughts About Drugs and Violence," p. 268.

52. Joan W. Moore, "Gangs, Drugs, and Violence," in *Drugs and Violence: Causes, Correlates, and Consequences*, ed. M. De La Rosa, E. Y. Lambert, and B. Gropper, NIDA Research Monograph 103, 1990, p. 169.

53. Cited with the authors' permission from John C. Quicker, Yvonne Nunley Galeai, and Akil Batani-Khalfani, "Bootstraps or Noose: Drugs in South Central Los Angeles," a draft report to the Social Science Research Council, March 1993.

54. Patrick J. Meehan and Patrick W. O'Carroll, "Gangs, Drugs, and Homicide in Los Angeles," *American Journal of Disease Control* 146 (1992): 683–87. I have yet to see any acknowledgment of these LAPD-inspired results from LAPD officials.

55. Dan Waldorf, "When the Crips Invaded San Francisco," unpublished paper, Institute for Scientific Analysis, Alameda, CA, January 1992, p. 152.

56. Jeffrey Fagan, "The Social Organization of Drug Use and Drug Dealing Among Urban Gangs," *Criminology* 27(4) (1989): 633–67.

57. Jeffrey Fagan, "Drug Selling and Licit Income in Distressed Neighborhoods: The Economic Lives of Street-Level Drug Users and Dealers," in *Drugs, Crime, and Urban Opportunity*, ed. S. Harrell and G. Petersen (Washington, D.C.: Urban Institute, in press).

58. Cited in Arnold P. Goldstein, *Delinquent Gangs: A Psychological Perspective* (Champaign, Ill.: Research Press, 1991), p. 211–12.

59. See the description by David Barton of the Kansas City Police Department, "The Kansas City Experience: 'Crack' Organized Crime Cooperative Task Force," in *The Police Chief*, January 1988, pp. 28ff.

60. The *Harbor Independent News*, San Pedro, Calif., November 12, 1992, reports that one drug ring may have "infiltrated" the LAPD and the district attorney's files, as noted in a court brief.

61. Scott Decker and Barik van Winkle, "Slinging Dope: The Role of Gangs and Gang Members in Drug Sales," draft paper, University of Missouri at St. Louis, November 1993.

62. Cited in *Gangs 2000: A Call to Action* (Sacramento: California Department of Justice, 1993) p. 20.

63. Isidor Chein, Eva Rosenfeld, and Daniel M. Wilner, *Studies in the Epidemiology of Drug Use: Progress Report on Third Study, Drug Use and Street Gangs* (New York: New York University, Research Center for Human Relations, 1954), p. 1.

64. Ibid., p. 6.

65. Victor Merina, "U.S. Counts on Felons in Trial of Drug Officers," *Los Angeles Times*, August 12, 1991, p. A15.

66. *Los Angeles Times*, April 14, 1992.

67. Felix Padilla, *The Gang as an American Enterprise* (New Brunswick, N.J.: Rutgers University Press, 1992).

68. The good old days of "the little old lady in tennis shoes from Pasadena" seem to have passed. The city has had a serious gang problem for decades.

Chapter 5

1. George W. Knox, *An Introduction to Gangs* (Berrien Springs, Mich.: Van de Vere, 1991), p. 247.

2. As Kobrin summarized this assumption, "Delinquency was seen as adaptive behavior on the part of the male children of rural migrants acting as members of adolescent peer groups in their efforts to find their way to meaningful and respected adult roles . . . under the influence of criminal models for whom the inner city areas furnish a haven." Solomon Kobrin, "The Chicago Area Project—A 25-Year Assessment," *Annals of the American Academy of Political and Social Science* 322 (March 1959): 23.

3. There is a striking and ironic parallel in this 1930s depiction and the philosophy underlying the "modern" development of community policing. The "new" conception that community residents should have a major voice in how police resources should be deployed is hardly new. Even its appeal as a new approach to gang control is merely a rediscovery of the spirit of the Chicago Area Projects.

4. Kobrin, "The Chicago Area Project," pp. 20–29; quotation from p. 28.

5. Daniel P. Moynihan, *Maximum Feasible Mis-Understanding: Community Action in the War on Poverty* (New York: Free Press, 1969).

6. This was deemed at the time to be an unthinkable number. Seen in today's perspective—4,600 gang-related deaths in Los Angeles County in the last ten years—it reminds us just how tied our views are to our own time context.

7. A brief summary of these developments is contained in the report "Dealing with the Conflict Gang in New York City," prepared in 1960 by the Juvenile Delinquency Evaluation Project of the City of New York, located at City College.

8. Robert Rice, "Who You Are and What You Think You're Doing." Reprinted by permission: © 1961, 1989, *The New Yorker*, Inc. First printed December 23, 1961, p. 32.

9. The reader interested in the nuts and bolts of street work operations should look at four publications in particular: (1) *Reaching the Fighting Gang*, New York City Youth Board, 1960; (2) James E. McCarthy and Joseph S. Barbaro, "Re-directing Teen-age Gangs," in *Reaching the Unreached*, New York City Youth Board, 1952; (3) Irving Spergel, *Street Gang Work: Theory and Practice* (Reading, Mass.: Addison-Wesley, 1966); and (4) Frank Carney, Hans W. Mattick, and John D. Callaway, *Action on the Streets* (New York: Association Press, 1969).

In addition, Spergel's forthcoming book for Oxford University Press, *Youth Gangs: Problem and Response*, contains an excellent review of detached worker programs.

10. McCarthy and Barbaro, "Redirecting Teen-age Gangs."

11. Ibid., p. 110.

12. William E. Amos, Raymond L. Manella, and Marilyn A. Southwell, *Action Programs for Delinquency Prevention* (Springfield, Ill.: Thomas, 1965), p. 60.

13. Carney et al., *Action on the Streets*, p. 15.

14. Ibid., p. 16.

15. But that's not one hundred members: Forty-five members received

these positions. Purple Pedro (see Chapter 3) went through ten of these jobs all by himself.

16. Twenty months after the project ended, inquiries into the status of the Latin members revealed a very mixed picture. Some were married and employed, and others were married or employed. Half a dozen had entered the armed services, and several had been wounded in Vietnam. A number had left the area; others were incarcerated. A few were still hanging around Ladino Hills with little to show for their efforts. A dozen were dead as the result of overdoses or homicides. It's not a pretty picture, overall.

17. The CYGS's crisis hotline phone number was 626-GANG!

18. See Dorothy Torres, *Gang Violence Reduction Project: Fourth Evaluation Report* (Sacramento: California Youth Authority, 1981).

19. See the reference and other pithy examples of the cop–gang member interaction in Carl C. Werthman, "Delinquency and Authority" (master's thesis, University of California at Berkeley, 1964).

20. *Dealing with the Conflict Gang in New York City*, pp. 2–3.

21. Los Angeles Sheriff Sherman Block, quoted by Louis Sahagan, "Fight Against Gangs Turns to Social Solution," *Los Angeles Times*, November 11, 1990, p. A3.

22. Louis Sahagun, "Sheriff's Officials, Community Leaders Shift Gang Strategy," *Los Angeles Times*, January 16, 1991, p. 33.

23. Quoted on p. 25 of the *Proceedings of the National Conference on Youth Gangs and Violent Juvenile Crime*, Reno, Nevada, 1991, available from the National Youth Gang Information Center, OJJDP.

24. *Gang Violence in Philadelphia* (Pennsylvania Crime Commission, July 1969) p. 27.

25. Readers wanting to organize community gang programs should first read Irving Spergel, "General Community Design for Dealing with the Youth Gang Problem," revised draft (Chicago: University of Chicago School of Social Service Administration, August 1990), pp. 66–67.

26. A 250-page manual for adoption elsewhere is "Rising Above Gangs and Drugs: How to Start a Community Reclamation Project," available from the CRP in Lomita, California.

27. Barry Krisberg, *The Gang and the Community* (Berkeley: University of California School of Criminology, 1975).

28. From the Life and Arts section of the *Daily Trojan*, University of Southern California, September 17, 1990.

Chapter 6

1. Spergel, in turn, pulled off quite a trick. Believing, on the basis of his long experience, that suppression had value only in a broader context of community organization, he produced a suggested program of modules (police, prosecution, corrections, schools, community agencies, etc.), with each embedded in an overall community organization rubric. No "alternative" suppression models emerged. The OJJDP lost sight of its original intentions and reluctantly went along with Spergel's reinterpretation of the mandate. This, I freely admit, is my interpretation, based on my early discussions with the OJJDP staff well before Dr. Spergel became involved.

2. What do we know about crime crackdowns generally, since gang crackdowns have not been evaluated? Sherman's extensive review of eighteen crackdown case studies revealed a pattern of generally good initial success followed shortly by a decay in the effect, even when increased "dosage" was applied. Only five cases seemed genuinely effective, and each involved crackdowns on relatively minor offenses or middle-class forms of crime. The eighteen cases involved emphases on such offenses as illegal parking, drug sales, drunk driving, prostitution, subway crime, and cable TV fraud. No instances of gang suppression were assessed. But there may be transferable implications, as Sherman's review not only warns of effect decay but also cautions us to assess backfire, displacement, and a set of ethical issues as well. Hayeslip's review of drug enforcement suppression efforts refers to many of the forms of suppression to be found in our gang cities as well—crackdowns, civil abatement, asset seizure, saturation patrols, reverse stings, and street sweeps—but concludes that all one sees are "program outputs (actions of police) rather than program outcomes (reduction in crime)." Uchida and his colleagues' 1990 assessment in Oakland provided one of the few exceptions. In addition to their own project evaluation showing at least some positive results regarding citizens' perceptions and attitudes and mixed results regarding crime reduction, they reviewed an additional set of literature, including evaluations of stings, asset forfeiture, and "aggressive patrol." They concluded, "Overall, research on the effectiveness of drug enforcement strategies is still in its embryonic stage. Despite the great resources and creativity devoted by police to combatting street drug trafficking and drug abuse generally, knowledge about what works is limited in scope." Lawrence W. Sherman, "Police Crackdowns," *NIJ Reports*, 219 (March–April 1990): 2–6; D. W. Hayeslip Jr., "Local-Level Drug Enforcement: New Strategies," *NIJ Reports*, 213 (March–April 1989): 4–5; Craig Uchida, Brian Forst, and S. Annan, "Controlling Street-Level Drug Trafficking: Evidence from the Oakland Experiment," Police Foundation, report to the National Institute of Justice, Office of Justice Programs, U.S. Department of Justice, 1990, p. 2.

3. A major treatise on deterrence theory includes this trenchant comment about the simplistic assumptions of public officials who "seem to think in a straight line about the deterrent effect of sanctions; if penalties have a deterrent effect in one situation, they will have a deterrent effect in all; if some people are deterred by threats, then all will be deterred; if doubling a penalty produces an extra measure of deterrence, then trebling the penalty will do still better.... This style of thinking imagines a world in which armed robbery is in the same category as illegal parking, burglars think like district attorneys." Franklin E. Zimring and Gordon J. Hawkins, *Deterrence: The Legal Threat in Crime Control* (Chicago: University of Chicago Press, 1973), p. 19.

4. Mike Ward, "Crackdown on Gangs Pays off as Deaths Fall," *Los Angeles Times*, January 16, 1992.

5. Sweeps, like most suppression activities, are not new. Gonzales offers a description of Los Angeles's first gang sweep in 1942, the period of the Zoot Suit riots. See Alfredo Gonzales, "Mexicano/Chicano Gangs in Los Angeles: A Sociohistorical Cure Study" (Ph.D. diss., University of California at Berkeley, 1981).

6. *Los Angeles Times*, May 26, 1992.

7. Irving I. Spergel, *Youth Gangs: Problems and Response* (New York: Oxford

University Press, forthcoming), p. 184.

8. Joan W. Moore, *Going Down to the Barrio* (Philadelphia: Temple University Press, 1991), pp. 3–4.

9. Armando Morales, "Urban Gang Violence: A Psychosocial Crisis," in *Social Work: A Profession of Many Faces*, ed. Armando Morales and Bradford W. Sheafor, 5th ed. (Boston: Allyn & Bacon, 1989) pp. 413–50.

10. In contrast, I was invited to observe a coordinated set of raids on ten crack houses one night, carried out by the Los Angeles Sheriff's Department. A command post was set up several miles from the targets; a special communications system was established; the ten raids were coordinated in time; and all was carried out in secret with no media coverage. There was no "laughing material" available to the arrestees I observed that night.

11. This capacity for street gang members to turn negative experiences into positive values is what defeats so many of our attempts to intervene in gang affairs. In another context, Short also remarked on this capacity: "Defeat ...even fleeing the battlefield...offers status awards if the group has perfomerd with bravery...or the lack of clear-cut victory may be `explained' by the intervention of the police or a detached worker." See James F. Short, "Collective Behavior, Crime and Delinquency," in *Handbook of Criminology*, ed. Daniel Glaser (Chicago: Rand McNally, 1974), p. 412.

12. CRASH stands for Community Resources Against Street Hoodlums. I don't know where the *community* comes from. This is strictly a uniformed patrol operation. But it helps to know that the operation was originally labeled TRASH, Total Resources Against Street Hoodlums. Someone in the department had the sensitivity to suggest that TRASH was not a useful acronym in the inner city. Thus Total became Community.

13. This material was in a special article, "Policing Gangs; Case of Contrasting Styles," *Los Angeles Times*, January 19, 1986.

14. Ibid.

15. Robert J. Bursik and Harold G. Grasmick, *Neighborhoods and Crime: The Dimensions of Effective Community Control* (Lexington, Mass.: Lexington Books, 1993).

16. Report cited in *Criminal Justice Newsletter*, January 2, 1991.

17. *Los Angeles Times*, September 2, 1991.

18. Edward Burns and Thomas Deakin, "A New Investigative Approach to Youth Gangs," *FBI Law Enforcement Bulletin*, October 1989, p. 22. Deakin is a homicide detective in Baltimore.

19. See the extended treatment provided by Irving I. Spergel in his *Youth Gangs* and in *Youth Gang Activity and the Chicago Public Schools*, University of Chicago, School of Social Service Administration, 1985.

20. Prepared by the National School Safety Center (Malibu, Calif.: Pepperdine University, School Safety Center, 1988).

21. Associated Press coverage of the seventeenth National Conference on Juvenile Justice, March 28, 1990.

22. Ronald D. Stephens, "School-Based Interventions: Safety and Security," in *The Gang Intervention Handbook*, ed. Arnold P. Goldstein and C. Ronald Huff (Champaign, Ill.: Research Press, 1993), p. 219.

23. For a statisical summary, see Lisa D. Bastian and Bruce M. Taylor, *School Crime: A National Crime Victimization Survey Report* (Washington, D.C.: Bureau of Justice Assistance, September 1991). A very personal account by a

teacher and administrator in a troubled school is provided by Richard Arthur (with Edsel Erickson) in *Gangs and Schools* (Holmes Beach, Fla.: Learning Publications, 1992).

24. Paul Hoffman and Mark Silverstein, "Safe Streets Don't Require Lifting Rights," *Los Angeles Times*, March 11, 1993, editorial page.

25. The Blythe Street order was proposed by the Los Angeles city attorney shortly before his election to office. I think it is fair to speculate about the connection: Gang members seldom vote, whereas those threatened by their presence often do.

26. *Los Angeles Times*, September 20, 1989. Still, none of these examples is as extreme as the campaign proposal by a mayoral candidate to deport all members of two Latino gangs to Mexico and Central America. His source of information on the makeup of the two gangs in question was the U.S. Immigration and Naturalization Service (INS), which confirmed the next day that the two contained 23,000 members, "virtually all illegal immigrants" (*Los Angeles Times*, January 15, 1993). Of all the figures found in this book, this is undoubtedly the most ridiculous. What is it about federal agencies and street gangs?

27. The reader may want to review two interesting reports. The first is the Mitre Corporation's 1982 evaluation: Judith Dahmann, *An Evaluation of Operation Hardcore: A Prosecutorial Response to Violent Gang Criminality* (McLean, Va.: Mitre Corporation, 1982). The second is by Michael Genelin and Loren Naiman, *Prosecuting Gang Homicides*, published by the California District Attorney's Office, 1988, and available from the Los Angeles County district attorney. Genelin has directed the Los Angeles operation for many years.

28. *Gang Prosecution: Prosecutor Survey* (Alexandria, Va.: Institute for Law and Justice, June 1993).

29. See the *Program Policy and Procedure Handbook* (Los Angeles: Los Angeles County Probation Department, Specialized Gang Supervision Program, 1982).

30. See the more extended description in Malcolm W. Klein, "Attempting Gang Control by Suppression: The Misuse of Deterrence Principles," *Studies in Crime and Crime Prevention* 2 (1993): 88–111.

31. Jackson and Rudman's review of antigang legislation in California concludes, "The highly publicized triad of gangs, drugs and violence provided the major focus of virtually all the gang-related legislation generated during the 1980s." Patrick Jackson and Cary Rudman, "Moral Panic and the Response to Gangs in California," in *Gang: The Origins and Impact of Contemporary Youth Gangs in the United States*, ed. Scott Cummings and Daniel J. Monti (Albany: State University of New York Press, 1993), p. 261.

32. Two reports are available as this is written. One, a mere listing of statutes, was provided at my request. The second, "Gang Prosecution Legislation Review," is a useful and readable draft prepared for the National Institute of Justice. Copies are available from the National Institute for Law and Justice, 1018 Drake Avenue, Alexandria, Va.

33. The constitutionality of the STEP Act has been upheld by the appeals courts in at least two cases, *People* v. *Gomez* (1991) 235 Cal. App. 3d 957, and *People* v. *Green* (1991) 227 Cal. App. 3d 092.

34. Technical details can be found in the California Penal Code, section 186.22.

35. *C.Q. Researcher*, October 11, 1991, p. 761.

36. "LAPD `STEP' up Anti-Gang Activities," *Community Reclamation Project News.*

37. Phil Sneiderman, "8 Alleged Gang Members Arrested in Raid," *Los Angeles Times,* December 29, 1991, p. J1.

38. *People* v. *Green,* Vol. 227 Cal. App.3d 693; in re Nathaniel C., Vol. 228 Cal. App.3d. 990; in re Jose T. Vol. 230 Cal. App.3d. 1455.

39. Two other brief prescriptive sets of guidelines, documents D0144 and D0145, are available from the LA city attorney's Gang Prosecution Section or from the National Youth Gang Information Center: *Civil Gang Abatement* and *Building Abatements.*

40. An excellent review of the applicability of RICO to gang situations is provided by Robert A. Destro, "Gangs and Civil Rights," in *Gangs: The Origins and Impact of Contemporary Youth Gangs in the United States,* ed. Scott Cummings and David J. Monti (Albany: State University of New York Press, 1993). Destro also covers the issues in civil cases such as that of the Blythe Street gang and the closing of public parks to gang members.

41. Quoted in a 1992 newsletter of the Midwest Gang Investigators' Association.

42. In Minnesota, the law was ruled unconstitutional because it disproportionately affected blacks.

43. The first time it was applied, to a mother who had pictures taken of herself with her children and gang members brandishing various firearms, the charges were dropped because the vagueness of the law's language made successful prosecution unlikely. Besides, this mother was able to prove that she had been enrolled in a parenting class!

44. Jackson and Rudman, "Moral Panic and the Response to Gangs in California."

45. Joan Moore and Diego Vigil, "Chicano Gangs: Group Norms and Individual Factors Related to Adult Criminality," *Aztlan* 18 (1989): 31.

46. Zimring and Hawkins, *Deterrence.*

Chapter 7

1. John C. Quicker, *Seven Decades of Gangs* (Sacramento: California Commission on Crime Control and Violence Prevention, 1983), pp. 125–26.

2. See *"Los Angeles Style:" A Street Gang Manual of the Los Angeles County Sheriff's Department,* January 1992.

3. Helaine Olen and Paul Lieberman, "National Tracking of Gangs Proposed," *Los Angeles Times,* February 1, 1991, p. B1.

4. I fervently hope that the reader who has not come across this volume will put it next on the list. It was published in 1987 by the University of Chicago Press and has become a major focus of much recent theorizing about urban America.

5. In addition to *The Truly Disadvantaged,* Wilson has provided useful statements of the model in several sources. See his ASA presidential address, "Studying Inner-City Social Dislocations: The Challenge of Public Agenda Research," *American Sociological Review* 56 (1991): 1–14; "The Underclass: Issues, Perspectives, and Public Policy," *Annals of the American Academy of Political and Social Science* 501 (1989): 182–92; and "Public Policy Research and the Truly Disadvantaged," the concluding chapter in *The Urban Underclass,* ed. Christopher Jencks and Paul E. Peterson (Washington, D.C.: Brookings

Institution, 1991). In fact, the whole volume on underclass issues is worth the reader's attention. A useful summary of the issues is contained in Martha A. Gephart and Robert W. Pearson, "Contemporary Research on the Urban Underclass," *Items* 42 (1988): 1–10. This publication from the Social Science Research Council helped launch the council's special program on the urban underclass. Other useful statements are those by Troy Duster, "Crime, Youth Unemployment, and the Black Urban Underclass," *Crime and Delinquency* 33 (1987): 300–16; Douglas S. Massey and Mitchell L. Eggers, "The Ecology of Inequality: Minorities and the Concentration of Poverty, 1970–1980," *American Journal of Sociology* 95 (1990): 1153–88; and Joel A. Devine and James D. Wright, *The Greatest of Evils* (Hawthorne, N.Y.: Aldine de Gruyter, 1993).

6. Joan W. Moore, "Gangs, Drugs, and Violence," in *Drugs and Violence: Causes, Correlates, and Consequences,* ed. M. De La Rosa, E. Y. Lambert, and B. Gropper, NIDA Research Monograph 103, 1990, p. 172.

7. Jeffrey Fagan, "The Political Economy of Drug Dealing Among Urban Gangs" (draft), 1992, Rutgers University.

8. *The City in Crisis*, a report to the Board of Police Commissioners, October 21, 1992, p. 40.

9. Thomas J. Bernard, "Angry Aggression Among the Truly Disadvantaged," 28 *Criminology* (1990): 73–96.

10. Malcolm W. Klein, *Street Gangs and Street Workers* (Englewood Cliffs, N.J.: Prentice-Hall, 1971), p. 91.

11. See John Hagedorn's *People and Folks: Gangs, Crime, and the Underclass in a Rustbelt City* (Chicago: Lakeview Press, 1988); and two other articles: "Gangs, Neighborhoods, and Social Policy," *Social Problems* 38 (1991): 529–42; and "Back in the Field Again: Gang Research in the Nineties," in *Gangs in America*, ed. C. Ronald Huff (Newbury Park, Calif.: Sage, 1990).

12. Hagedorn, *People and Folks*, p. 128.

13. William Julius Wilson, "Imagine Life Without a Future," *Los Angeles Times*, May 6, 1992, p. B9.

14. Ibid.

15. Once again, I exclude the popular drug explanation for a variety of reasons:

1. Drugs as a potential cause has been around for many decades.
2. Most of this gang explosion began before the recent crack explosions.
3. As best we can determine, there are many gang cities without a major growth in crack and many crack cities without a major growth in gangs.

16. For example, Thomas J. Bernard, "Angry Aggression Among the `Truly Disadvantaged'," *Criminology* 28 (1990): 73–96, uses urban underclass variables to explain the development of "angry aggression" as a subcultural phenomenon (with no reference to gangs).

17. Nathan L. Gerard, "The Core Member of the Gang," *British Journal of Criminology* 11 (1964): 361–71.

18. Klein, *Street Gangs and Street Workers*, p. 83.

19. Herbert C. Covey, Scott Menard, and Robert J. Franzese, *Juvenile Gangs* (Chicago: Thomas, 1992), p. 178.

20. C. Ronald Huff, "Youth Gangs and Public Policy," *Crime and Delinquency* 35 (1989): 524–37.

21. Pamela Irving Jackson, "Crime, Youth Gangs, and Urban Transition: The Social Dislocation of Postindustrial Economic Development," *Justice Quarterly* 8 (1991): 379–97.

22. David Curry and Irving I. Spergel's paper, "Measuring Gang Involvement Among Hispanic and Black Adolescent Males," is available from David Curry at West Virginia University in Morgantown, West Virginia (draft, 1993).

23. From the conference summary, *Ohio Conference on Youth Gangs and the Urban Underclass*, prepared by C. Ronald Huff (Columbus: Ohio State University, 1988). A balanced summary is offered by Scott Cummings and David J. Monti in the last chapter of their book *Gangs: The Origins and Impact of Contemporary Youth Gangs in the United States* (Albany: State University of New York, 1993), p. 307:

> Clearly, wholesale application of Wilson's version of underclass theory to describe the history, growth, and development of gangs in American cities has limitations. We know, for example, that gangs are not exclusively drawn from minority populations. We know that gangs are not exclusively comprised of young people drawn from the very lowest strata of society. We also know that many gangs have long-standing cultural traditions, and are deeply rooted in the fabric of community life and culture. Having acknowledged these limitations, we are also convinced that Wilson's treatment of the emergence of a permanent underclass in American cities has important ramifications for development of effective public policies dealing with gangs.

24. I'm indebted to Terri Moffitt for this parallel, taken from a draft paper to appear in the *Psychological Review*.

25. Anne Campbell, Steven Munce, and John Galea, "American Gangs and British Subcultures: A Comparison," *International Journal of Offender Therapy and Comparative Criminology* 26 (1982): 76–89.

26. From an interview reported in *Connections*, the Family and Youth Services Bureau newsletter, Summer 1991.

27. Sacramento: *OCJP* 11 (summer 1990): 11.

28. England provides an apt case. "Thus, public discussions and portrayals in the media about gang rivalries...acted to create gang rivalries...and the structure and purpose of potential gangs. Writing about two groups in a similar situation in England, Stanley Cohen writes: `Constant repetitions of the warring gangs' image...had the effect of giving these loose collectives a structure they never possessed and a mythology with which to justify the structure.'" Alfredo Guerra Gonzales, "Mexicano/Chicano Gangs in Los Angeles: A Socio-historical Case Study" (Ph.D. diss., University of California at Los Angeles, 1981), pp. 121–22.

29. The same source, the Long Beach police, places the cost to the county at $12 million, to the state at $29 million, and to the nation at $4.5 billion to remove graffiti.

30. John P. Sullivan and Martin A. Silverstein, "The Disaster Within Us: Urban Conflict and Street Gang Violence in Los Angeles" (paper prepared for an international conference in Stockholm, 1993).

31. WST—West Side Trouble—may be an exception. An estimate of unknown reliability suggests that there are five hundred WST taggers.

Chapter 8

1. Lois DeFleur, "Delinquent Gangs in Cross-Cultural Perspective: The Case of Cordoba," *Journal of Research in Crime and Delinquency* 4 (1967): 132–41.

2. Jean Monod, "Juvenile Gangs in Paris: Toward a Structural Analysis," *Journal of Research in Crime and Delinquency* 4 (1967): 142–65.

3. T. R. Fyvel, *Troublemakers: Rebellious Youth in an Affluent Society* (New York: Schocken Books, 1964).

4. Marshall B. Clinard and Daniel J. Abbott, *Crime in Developing Countries: A Comparative Perspective* (New York: Wiley, 1973).

5. Irving I. Spergel's list comes from his *Youth Gangs: Problems and Response* (New York: Oxford University Press, forthcoming). The other is the review text by Herbert C. Covey, Scott Menard, and Robert J. Franzese, *Juvenile Gangs* (Chicago: Thomas, 1992).

6. "Les Casseurs," *Le Nouvel Observateur*, November 22–28, 1990.

7. A dramatic portrayal can be found in Luis Francia, "The Dusty Realm of Bagong Barrio," *Icarus* 3 (summer 1991): 13–30.

8. Wu Zai-de and Xu De-qi, "The Longitudinal Analysis of the Shaping and Growth of the Heads of Teenage Gangs" (paper delivered at the NATO Workshop on Cross-National Longitudinal Research on Human Development and Criminal Behaviors, Freudenstadt, Germany, July 1992).

9. Or at least now puts "gang" in quotation marks!

10. The person who devised the clique analysis (not unlike that in my Figure 3-1) had taken a university class with Sarnecki.

11. The visiting voyeur should also check out Kungsgatan, Levinsky's Cafe near Stureplan, the T-Centrallen Station, and the McDonalds on Sveavagen.

12. Monica Olsson gave me some preliminary data, but more are yet to come.

13. The sellers are deliberately obvious: mustachioed Moroccans posted at intersections, park entrances, and subway outlets.

14. In England, an "estate" is a housing development, not a mansion and grounds, as in the American usage.

15. See Marie C. Douglas, "Auslander Raus! Nazis Raus! An Observation of German Skins and Jugendgangen." *International Journal of Comparative and Applied Criminal Justice* 16 (1992): 129–34.

Index